LIVING ON THE THIRD PLANET

Hannes and Kerstin Alfvén

Translated by Eric Johnson

W. H. Freeman and Company
San Francisco

Contents

Preface vii

I The Condition of Life 1

II The Race from the Past to the Future 25

III Life and the Cosmos 39

IV Man's New Environment 53

V Smoke Screens 67

VI The Politicians 79

VII Man's Increasing Ignorance 101

VIII The Scientists 119

IX The Population Explosion 137

X The Symbiosis Between Man and Technology 151

XI Are We Unique? 165

Preface

On the third of the planets orbiting our sun, life has developed. In its cosmical, physical, biological, and social aspects, life is an extremely complicated phenomenon. A large number of specialists have studied it in detail, but not even all the books they have written suffice to describe it completely. Moreover, the specialists often have controversial views.

A complete description of life on Earth is of course impossible, particularly in a book of about 200 pages. Moreover, such an attempt is necessarily an offense to all specialists, because the authors are certain to be amateurs in all the fields the book should cover, except those in which they themselves specialize. Nevertheless, in an age when specialists have become so specialized that they fail to communicate with each other, we feel it is important to extract from various disciplines some of the specialists' knowledge, and to try to put it into a consistent pattern. Such a synthesis is essential if we are to discuss and evaluate the strategic position of mankind. For this purpose we stress the long-term perspectives, avoiding as much as possible those issues that are likely to be forgotten within a decade or two. But our real purpose in this book is nothing more than to take a snapshot of life on our planet and put it into an appropriate cosmic frame.

Most of the facts quoted in the text are commonly known, at least to students of science. In the appendix "References and Comments," we list some sources we have used and discuss a few controversial issues in more detail.

HANNES ALFVÉN
KERSTIN ALFVÉN

September, 1971
La Jolla, California

LIVING ON THE
THIRD PLANET

I

The Condition of Life

Four or five billion years ago an enormous amount of hydrogen was formed into a giant ball, a ball which we today call the sun. This formation was a result of gravitation, which condensed and heated the matter. When the temperature in the center of the sun reached 15 or 20 billion degrees, nuclear reactions began transforming hydrogen into helium in a giant fusion reactor, which radiated its energy out into space. For several billion years the sun has remained at about the same temperature and will continue to do so for several billion years to come.

At the same time that the sun was formed, a number of planets were produced around it. Life originated on the third planet. Many people have speculated about life on other planets, and imagination has populated Venus and especially Mars. But such speculations have been disproved by facts: for one of the major results of space research has been the recognition that it is highly improbable that life exists on any other celestial body in the neighborhood of the sun. We are alone; we are unique, at least in this region of space. If we have some fellows, they are many light years away from us.

When the earth originated, it was sterile. Only after some time did it, like a ripening cheese, become moldy. We do not know in detail how life originated, but we can reconstruct the major occurrences in the causal processes. Through the action of cosmic forces, solar radiation, and perhaps also electric discharges, large quantities of very complicated carbon compounds were produced. The carbon atoms have a remarkable capability to combine into long chains or rings, which contain other elements, such as hydrogen and oxygen. Almost all elements—except the noble (or inert) gases—can be incorporated into the more complicated atomic structures. In the course of millions and millions of years, the most incredibly complicated structures were formed, and these in turn disintegrated and transformed into other structures through continued processes. The flow of energy producing these changes came from solar radiation, which originated from the giant fusion reactor at the center of the sun.

Perhaps on one single occasion, or perhaps many times, a giant molecule was produced that had the primary property of life, the ability to reproduce itself. It could selectively absorb certain substances from its environment and use them for building another giant molecule of identical structure. The new molecule could live in the same way on the substances of its surroundings, and it too could produce new molecules of its own kind again and again.

LIFE AS A SERIES OF POPULATION EXPLOSIONS

In the foregoing way the first population explosion occurred. All obtainable food was eaten up by the hungry primitive being, which multiplied until lack of nourishment put a limit to the expansion of its numbers.

It is possible that the whole colony starved to death. But it is also possible that new food was produced in the same place by the same processes that had been active earlier. If this indeed occurred, the colony could have continued to exist. Perhaps some of these primitive beings were carried away by chance to another region where there was more food, thus allowing a new population explosion to take place.

Life, then, originated in this way. A special type of molecular compound, capable of self-reproduction, rapidly multiplied until it became quite common. But occasionally a reproduction was not exactly similar to the original compound; a "mutation"—or heritable alteration—took place. As a result of such mutations, a number of different species originated. After a long time a highly complicated structure called the cell was produced. Later many cells combined to form multicellular plants and animals. But life preserved its initial tendency—population explosion leading to death, or almost to death.

The same tendency is still inherent in all life. A piece of cheese gets moldy because a mold spore has happened to reach it and starts to multiply very rapidly. When all food—that is, the piece of cheese—has been changed into mold, the whole colony of mold dies. The mold as a species can continue to exist only if in this case—or in similar ones—a few spores of mold in one way or another have been able to reach another piece of cheese or something else that mold can feed upon, and thereby start another population explosion. Similarly, a germ which has happened to land in the mucus of a man multiplies rapidly until medicine or the body's protective mechanism has killed the whole colony. This particular species of germ can continue to live only if a few of the members of the colony—or similar colonies—have been able to infest another man.

A complicated animal (for example, a mammal) is in essence a huge colony of cells, and the same law holds true for those cells. At the moment a spermatozoon penetrates an egg cell, a population explosion of cells starts. The fertilized egg begins to divide, and the number of cells increases at an enormous rate. Billions of cells are produced. Marvelous processes, regulated by the chromosomes, cause these cells to interact and differentiate. The cellular structures in turn form a complicated being, an organization of many billions of cells. But this colony can increase only to a certain limit. It can keep its form and its functions for a number of years, but finally it will die. Almost all of the cells which have been formed by the fertilized egg are doomed to die within a limited time. Only a few egg cells or spermatozoa, which can continue to live and produce a colony of cells, will remain.

THE WELFARE STATE OF THE CELLS

When a group of cells form a multicellular animal, they become differentiated when they begin to interact. In the more complicated animals, the cells constitute a well-organized community. Instead of living a free but dangerous life, they thrive under the protection of the skin in a welfare state of cells. The alimentary organs function as the central sustaining force of the community, for they transform the substances that have been brought in from the surroundings into a nourishing solution which is distributed and delivered to all community members. In warm-blooded animals, the cells live in a house regulated by a thermostat.

The cells get what they need to live, but are in a state of subordination; they have lost much of their liberty and individuality. They must submit to differentiation: some are employed in the alimentary factory; others, in the distribution and delivery force. Still others are specialized into nerve cells, which respond to reports about the state of the organism's various parts, and which also bear responsibility for coordination and administration.

Though the nervous system governs the body, its power is nevertheless limited, because a number of bodily functions are regulated by hormones or by other chemical substances. Furthermore, there is not one nervous system but several, which are partially independent of each other. The alimentary processes of circulation, respiration, and all other "domestic and civil" functions are regulated by the vegetative nervous system in collaboration with the hormones.

Parts of the central nervous system operate as what we might call a military-industrial complex. This complex has the function of providing the whole society of cells with a maximum ability to protect the organism, to combat its enemies, and to seize food. It controls the arms and legs, and it can coordinate the motions necessary for attack in order to secure prey, or escape in case of danger. It can mobilize the civil administration for strong effort by the secretion of adrenalin. But the military administration does not have all the power: the civil administration can protect itself from overwork by closing down the military complex. This is what happens when we fall asleep or become unconscious.

This society ruthlessly sacrifices many of its inhabitants. In fact, the outermost layer of skin consists essentially of cell corpses.

If the complicated functions of the society become disorganized, every cell dies, because each cell must live under the laws of the welfare community. And each cell can live only under its protection.

LIFE AND DEATH

The requisite for life is the avoidance of death: the death of the individual, and the death of the species. No highly developed species can exist if it lacks either the instinct for self-preservation—consisting of hunger, fear, and other self-protective sensations; or the reproductive instinct—sexuality and child care. These properties have progressively evolved in the course of millions of years. Their great importance is manifested by the fact that eating and sexual intercourse have become two of the greatest pleasures of life.

But death always looms in the neighborhood of life. To live is to change, and the ultimate change is death. Death, however, can be postponed, since life can be stabilized to a certain extent. Whereas most of the unicellular animals die within an hour or a couple of days, the more complicated beings can live for perhaps ten to one hundred years, and a few can enjoy a relatively stable existence for an even longer time. The life of the species itself can be extended for many thousands of generations, and can remain reasonably constant for ten thousand years, perhaps even for one hundred thousand.

The number of individuals in a species varies. For example, the life of a colony of locusts consists of a number of population explosions followed by mass death, a cycle quite similar to that of the mold. But many other species enjoy a reasonably stable existence. In these species, the number of individuals increases during good years and decreases during bad years, but within rather narrow limits: the uninhibited multiplication, so disastrous to the locusts, has been stopped in some way.

In spite of the fact that reproduction is essential for the existence of a given species, it must be inhibited. A population explosion that is not stopped in time necessarily leads to mass death. This is a simple mathematical theorem.

THE EXPONENTIAL INCREASE IN POPULATION

Suppose that a colony consists of one hundred individuals which produce offspring, and that one hundred of these offspring reach an age at which they themselves reproduce. In such a case the number of individuals will remain constant through successive generations. But if we suppose that every group of one hundred produces two hundred offspring, the population will naturally double in every generation. At this rate, after ten generations the population will have increased more than one thousandfold, and after thirty generations it will have multiplied by as much as one billion. If we proceed further, we arrive at even more unreasonable figures. Mathematicians call this type of expansion an exponential increase. The lifetime of a species is often thousands of generations, but if it doubles itself for every generation it can continue to do so for only ten generations at the most. If the population explosion continues, it is a mathematical certainty that mass death will result. This is the essence of Malthusianism, a theory proposed by the British economist Thomas Malthus at about the turn of the nineteenth century.

Let us, as a mathematical example, consider a span of one thousand generations in the life of an imaginary species. Suppose that the species multiplies very slowly, so that the population of each generation is only one percent larger than that of the preceding one. In spite of such a slow rate of increase, after a thousand generations there will be several thousand times more individuals than there were in the first generation. Since even this rate of increase is one that continues unchecked, we can conclude that on the average not even a rate of one percent per generation is admissible if the species is to endure. Consequently it is certain that an increase in population of ten to one hundred per cent per generation cannot possibly continue for more than a few generations. We cannot escape the fact that, if such an increase should go on for long, the result would be certain mass death. Thus no species can long survive unless its average rate of population increase is not much more than zero for each succeeding generation.

In the long run, life and death must balance one another on the scale of collective survival. Thus, on an average, for every new individual that is born, an old one must die. Of course, a short-term exemption from this law is possible. The scale of life can outbalance the scale of death for a few generations, causing the species to increase rapidly. But very soon the equilibrium must be restored. When a species approaches its end, the scale of death outweighs the scale of life.

Every summer a maple tree produces drifts of seed which blow around. During its lifetime of perhaps a hundred years, it produces perhaps millions of seeds, but of all of them only a single one, on the average, is destined to grow into a similar big maple. If, in the next generation, an average of two new maples were to be produced from the seed of each tree, the number of maple trees would double. If this rate were to continue, the world would have space for nothing but maples within a few thousand years. At the other extreme, if only every second maple were to produce a new tree, the species would disappear in a few thousand years.

The foregoing is merely simple mathematics. Consequently every surviving species has two main alternatives. Its ratio of reproduction to death may be roughly constant at one to one, without much variation from that figure, so that the existence of the species remains stable; or it can follow the pattern of population explosion, which leads to mass death. Some of the more complicated animals follow the second pattern: that is, they multiply rapidly, and mass death subsequently occurs. But many species, notably most of the mammals and most of the plants, have rather stabilized existences. For these species, the reproduction rate changes very little: sometimes it is slightly above, sometimes slightly below, the ratio of one to one.

Because of the present explosion of the human population, which of the two groups man belongs to is uncertain.

LIFE FROM THE OUTSIDE AND LIFE FROM THE INSIDE

The expression "If you want to study a living thing, then drive out the spirit in it first" is, according to Goethe, the tradition of the life sciences. It is evident that life as seen from the outside looks completely different from life viewed from the inside. A butcher who is going to slaughter an ox evaluates it in terms of the current price of beef. We can hardly know how the ox looks on life, but we can assume that his philosophical outlook is considerably different from that of the butcher. Similarly, the statistician thinks of the people as an indeterminate number of social security numbers with associated data information. For the politician, they are also taxpayers and electors. But from the point of view of the people themselves, other factors are more important. In fact, "people" as such may cease to exist, to be replaced by a number of individuals. Each individual is his own center in his own world. In each of these worlds, the statisticians and the tax collectors and the whole mighty establishment are far out on the periphery, even if, from that peripheral basis, they sometimes make unpleasant attacks toward the center. Life from the outside, then, is significantly different from life from the inside: our world is dualistic. But this quality is not at all unique to our world. Even the atomic world is dualistic. We can consider an atom either as a planetary system of electrons, or as a resonance field of electron waves. These two views, the particle view and the wave view, are complementary. Seemingly, they are completely different, but in reality they never come into conflict. The uncertainty principle always enters as a protector. This principle states that it is impossible to ascertain simultaneously the position and the momentum of a particle, because the more accurately one can be measured the less accurately the other can be determined. It allows us to study the atomic world in two different ways, thereby permitting us to consider the atom from two different aspects.

Life is analogous to the atomic world, being also dualistic. When a person examines life from the inside—that is, his own life, his own world of thought—he does so in a way that is completely different from the way he views the lives of other people: he himself is the subject who reacts, feels, suffers, enjoys, acts, whereas other people are the objects that he loves or hates, and whose actions he sees. We all know, theoretically, that other people have the same soul and inner life as we ourselves possess, but it is often difficult to comprehend it. We may agree that we should love our neighbors, but it is too big a task for any individual to identify himself with everybody, to understand deep in his heart that everyone else faces problems of life as difficult as his own.

These dual perspectives on life, from within and without, are fundamental to life. The problems of human consciousness, freedom of will, and the moral responsibility of man are different aspects of this dualism.

Attempts have been made to correlate the dualism of life with the dualism of the atomic world. Some people have tried to make use of the uncertainty principle to explain the freedom of will. Such reasoning is probably faulty. To be sure, there are similarities between the two dualisms, but the analogy is only superficial.

The dualism of life allows us to approach the mystery of life—that incredible complication of life—in two different ways. One is the religious, philosophical, humanistic, artistic way—the way from the inside; the other is the statistical, biological, evolutionistic way—the way from the outside. Both approaches are equally important. Just as we must accept the wave-particle dualism in the atomic theory, so we must understand life itself from two aspects simultaneously.

In the development of the atomic theory, scientists discovered the particle nature of the atom first and later its wave nature, subsequently synthesizing the two. The whole series of discoveries took place in the course of only one or two decades. But the atom is a very simple structure; the whole of atomic physics is elementary in comparison with the complexity of the problems encountered in the study of life. This attempt to understand life is a gigantic undertaking, one that has barely begun. Furthermore there is no guarantee that man will ever be able to understand it completely.

The first attempt to analyze life was an approach from the inside. At the time of the earliest recorded human history, this problem was in the center of early theoretical investigations. The great religions which evolved between two and four thousand years ago were an expression of the results of man's introspection. The human condition, the possibility for the human soul to find peace, the ethical aspects of human interrelations had already at that time been analyzed with an acumen and depth of insight which can hardly be surpassed today. Indian literature, in particular, is impressive. The psychic exercises of yoga, the rigid asceticism, the sensuality of the classic Hindu treatise on love—the *Kama Sutra*—and the meditation and the logical stringency of the religious speculation all give an idea of the spectrum of experience one can attain through adherence to these beliefs. Another example of these early introspective analyses is the ancient Greek philosophy which flourished two millennia ago. The Greeks, too, grappled with the fundamental questions of life, but of course from a somewhat different perspective.

By way of contrast, the study of life from the outside is made through such disciplines as statistics, sociology, and neural physiology. Of special interest is the fact that scientists have begun to analyze which neurological processes cause different states of mind. Soon we will have the ability to analyze human thought from a purely scientific point of view. But what we have thus far understood is no more than a humble beginning. The study of life from within has a head start of more than two thousand years, and it will take a long time for science, with its approach from the outside, to catch up. And not until it has caught up will we be able to acquire a dualistic understanding of life.

THE AMOEBA SCHEHERAZADE

We have discussed the exponential growth of a given colony of individuals as it is viewed from the outside, but how is it experienced from the inside? Let us make a hypothetical, if unreasonable, assumption that an amoeba is able to understand itself and its situation in life. Also, for the sake of simplification, let us make the further assumption that an amoeba which has plenty of food will divide once a day. This means that one single amoeba in one year will divide into 2^{365} which is more than 10^{100} (a number with more than 100 figures). The balance between life and death must be established quite rapidly. This means that, on the average, the number of deaths must be equal to the number of births. Now let us read what the hypothetical amoeba Scheherazade tells us about this growth: "I originated from a certain mutation several million years ago. In reality I could trace my ancestry still farther back, in fact, back to the origin of life. I am exactly like all other amoebas, and yet there is something unique about me. All others have died, or almost all others have; whenever I met another amoeba she was usually dead within a day or two. Of all my friends, there was only one in a billion who survived one month, and one in a billion billions who survived two months. But I seem to have been singled out by divine grace. I was saved in the most incredible, remarkable ways from all accidents which killed the others. There must be some God who has protected me in such a way; or perhaps some destiny. I am beyond doubt one of the elect. I am immortal. Consequently I have a fatalistic view, almost a religious attitude towards life. But there seem to be a few more besides myself who enjoy the same privilege of immortality. I have discussed my life with some of these other amoebas and it is evident that all of them feel that they are chosen. They have all lived millions of years, and all have been saved from innumerable catastrophes. Therefore, I feel sure that we all are immortal." When Scheherazade had finished this story, she died.

There was no cruel sultan who decapitated her. All that happened was that she was entered in the biggest column of statistics, the column of death.

At least part of what Scheherazade said was true. Of the amoebas that don't divide, half are dead the next day and only one in a billion is still alive after one month.

Even of those that do divide, not many that are alive today will have progeny that will still be alive within one month. Millions of years of the most fantastic escapes do not give any guarantee of immortality. On the contrary, almost every amoeba that is alive now lives very close to her death. Her past extends for millions of years, but her future, probably only a couple of days: simple statistics, but seen from the inside.

Is the situation of mankind similar to that of the amoeba Scheherazade? Man has become the master of nature. But does this mean that he is no longer subject to the general conditions of life? This is what we shall discuss in the following chapters.

REGULATING THE POPULATION

The exponential increase of the population is a mathematical model and, like many such models, it informs us of something essential, in this case one of the conditions of life. But life is much too complicated, much too wonderful to be fully captured and understood by simple models.

In this section, we shall discuss some typical mechanisms for regulating the population. The number of animals of a certain colony depends often upon the balance between food and enemies. For example, if the hares in a certain region find plenty of food, they multiply exponentially as long as the food supply is sufficient. But at the same time hares are food for foxes, and the more hares there are the more food the foxes have, so that they too are able to increase exponentially. The result is a complicated interaction between hares and hare food and foxes and fox food, which in turn limits both the number of hares and the number of foxes. This example illustrates the two most important mechanisms that regulate a population: starvation and violent death. A third mechanism is of course disease. For man, and for a very few other species, another is war. However, in actuality, the size of a population is usually regulated in a much more complicated way.

Many biological investigations have been made in order to study how a group develops when it is allowed to multiply. Some of these investigations have been made because they are relevant to the situation of mankind. In one such study, a few rats of both sexes have been locked in a spacious cage and have been provided with plenty of water and food so that they have every reason to find life enjoyable. The cage is divided into a number of rooms with free passage between them. In the beginning the rats multiply very rapidly. A female usually breeds litters of five to ten offspring, and since there is plenty of food most of them grow into mature rats that reproduce. In the beginning everything is peaceful and harmonious. But when the population increases, the behavior of the rats begins to change. Fights between the males become more and more common, especially around the food stores, even though there is food enough for everybody. But these conflicts do not result in death. Very few animals, with the exception of man, kill members of their own species. In a colony of rats, something like a class society begins to develop. Those rats which are victorious in the fights occupy the best part of the cage and surround themselves with females. They then defend their position against the less aggressive rats. But the females get more and more annoyed by the crowding, becoming especially disturbed when they are nursing their offspring. As a result they neglect their litters, and sometimes even eat them. The close crowding of the rats thus leads to a change in their general behavior. They appear to be under heavy stress and become increasingly aggressive toward each other. As a further consequence, the number of newborn diminishes and the mortality rate of the young increases rapidly. An equilibrium is finally attained. Thus in spite of the fact that there are unlimited quantities of food, the number of rats does not increase. The limiting factor seems to be that their society has become too unpleasant, too full of tension and stress, owing to the high population density. This is a cruel experiment, almost as cruel as the experiment which mankind, through its own population explosion, is carrying out on itself.

Other biological investigations show that some birds have an intricate social mechanism that regulates their number. Among the thrushes, for example, every male occupies a certain region. He sits at the top of a tree and sings to announce his domination, and he tolerates no other thrush in the immediate vicinity. Because the region he occupies is big enough to feed his family, no member of his family needs to be hungry. Should there be too many thrushes, a number of them will not be able to live within the larger surrounding area. These must then emigrate to a less favorable region in which it is more difficult to find a female and build a nest. Therefore they produce fewer offspring. These investigations show, then, that there is probably a social mechanism that regulates the thrush population, and that bird song is an important factor in that mechanism. Such a self-regulating mechanism seems more humane than the pattern in which a population explosion leads to mass death.

LIFE IN A LAKE

We have discussed some mechanisms which check the population explosion of certain species. To treat population explosion in general, we have to study ecology—the interaction between all the species in a certain region, how they multiply and interfere with each other, how they consume the available food and produce waste, how in different ways they establish a coexistence, an equilibrium.

We have seen how cells may collaborate with each other and form a welfare state, an organism which may be a plant or an animal. But the process of the integration of a number of small living units to form a larger colony does not stop at that stage, although the next step has a different character; all the plants and animals living in the same region interact with each other in such a way that the life of each individual to some extent affects the life of everybody else. They form what is called an ecological unit: each of them constitutes a part of the milieu of the other. A lake, a wood, a prairie: all constitute ecological systems. Such systems may change with time—trees may begin to grow on a prairie, changing it to a forest—but often a region retains its character for a long time, remaining populated for many years or centuries by the same kinds of species in approximately the same numbers.

Some small lakes are typical examples of an ecological system, because biologically their inhabitants do not interact very much with the surrounding area. In a lake a complicated interaction takes place between a number of plants, animals, and microorganisms. The plants constitute the food of insects, larvae, and small fish, which are themselves the food of big fish. When the plants and the animals die, they are all food for a large variety of microorganisms, which make from the corpses new chemical compounds, some of which are fed again and again into the life cycle of the lake.

All the biological processes take place in water that is supplied to the lake by rivers or springs, which also supply mineral salts, sand, and mud. In addition, the oxygen and carbon dioxide which life requires are at least partly supplied by absorption of air at the surface. The energy that is necessary to keep all the processes going comes from the sunlight which shines on the lake.

The state of the lake changes with the time of the year. It may freeze in the winter, the water level may rise in the spring, and the lake may be close to drying up in the summer. These conditions affect the life in the lake, and only species who can stand this yearly cycle can survive. This type of fluctuation is one of the mechanisms that selects the possible species and determines the ecology of the lake.

In the lake society there is collaboration among the members, but there is also competition, a fight for survival, between individuals as well as between species. But under certain conditions this competition results in what seems to be an equilibrium: the lake can be very stable in the sense that the same species live there for centuries or millennia. The number of individuals in each species varies from one year to another, but may, on the average, be quite constant for very long periods.

To some extent the whole ecological system of the lake can be considered as a giant organism. In most organisms, animals or plants, cells are born and die, but the whole organism may continue its life during a period which is much longer than the lifetime of an individual cell. In a lake every individual may die within a span of a few years, but the species and the whole "ecosystem" may endure for many centuries.

Nevertheless the life span of a lake is finite. All small lakes which exist now are young from a geological point of view; that is, they are less than a million years old. The causes of death for a lake are similar to those for man: an external catastrophe, loss of its means of subsistence, old age.

If the geological structure of the region in which the lake is situated is changed, the rivers which feed water into the lake may change course, so that the lake dries out. Or the climate may change so that not enough rain falls in the region which drains into the lake. Then the lake region may change into a desert, and nothing is left of the lake except a number of geological deposits from which a geologist can make a post mortem and reconstruct the life history of the lake.

Today many lakes are killed by man, by irrigation, and very often by pollution. But we shall not discuss this in the present chapter. Instead, let us consider a lake which existed several million years before *Homo pollutans,* The Great Pollutor, began his work. Even at that time the normal cause of death of a lake was "pollution"; but the pollution was caused by the lake life itself (mostly by the plant life). Although the ecosystem of the lake seemed to be in equilibrium for centuries or even millennia, a slow evolution was constantly taking place in such a way that life in the lake was necessarily self-destructive. This is analogous to the processes in the bodies of us all: when the life span of a man is at an end, the biological processes in his body are self-destructive.

The character of a lake varies enormously, depending on climate and geological conditions. A particular lake may have been produced by geological forces which formed a basin in which water accumulated, or it may have been caused by a change in the climate, such as the one that occurred in northern Europe when the ice melted at the end of the Ice Age. Originally the lake may have been a basin of clear water, with a bottom of clean sand and sparse vegetation. Then a few plants grew there, a few fishes fed on them, and a few microorganisms worked to reprocess the organic material. This was the pioneer stage, when life invaded an uninhabited, almost virginal region. But not all of the organic material was recirculated. A quantity of it was deposited on the bottom, mud was produced, and the first result was that the lake became more favorable for organic life. The pioneers expanded: the vegetation slowly became more abundant, and more animals were bred. If looked upon with a time-scale of decades or centuries the ecology seemed to be in equilibrium, but this "equilibrium" was gradually changed. More and more mud accumulated at the bottom, the vegetation became denser, perhaps the oxygen content became lower, and "slum" areas were produced, with high population density and a scarcity of food. Ultimately the lake changed into a marsh, and later perhaps into a wood or a prairie. A few fishes and other organisms may have escaped death by leaving the lake through its outlet. The rest of their colonies died. A new ecosystem, perhaps with prairie flowers and grazing cattle, invaded the region.

Why did life in the lake kill itself? Did it not have the preservation instinct which life usually possesses? The answer is no. All animals and plants have certain mechanisms to preserve their individual lives, and every species has a mechanism for continuing the life of the species. But an ecosystem does not have this property.

Every individual, be it plant or animal, fights for its survival, taking food wherever it can. Every species insures its survival by breeding new individuals as fast as it can. There are a number of autonomous processes of this kind, which collaborate when the lake is young. But the eventual result of all this activity is the gradual deterioration of the general conditions. The essentials for life become more scant, the waste products more choking. It is, then, the desperate fight for individual survival, and for the survival of the species, that causes the conditions for life to deteriorate. The lust for life of everybody means the death of everybody.

Suppose there was an ecologist who understood all these processes, and who wanted to save the lake. He would recognize that if the different species were a little less ambitious, they could save themselves and everybody else. He could give them the following advice: "Slow down a little, and do not reproduce at such a rapid rate. The resources that are necessary for life are not abundant enough for the present rate of expansion, and your own waste will choke you. You cannot go on depositing so much mud on the bottom." But none of the individuals, none of the species, would heed him. They would all continue to fight for life, in a way that meant eventual death.

The biosphere—all life on our planet—forms an ecological system much more complicated than the ecosystem of a small lake. But perhaps the general laws of life are the same. Whether we recognize these laws is decisive for the future of man. Will our species be doomed, like all the species in the lake, or will we survive? If so, for how long? Before we discuss this we must look at the position of man in the biosphere.

II

The Race from the Past to the Future

MAN'S HISTORY

Today man is without any doubt the master of the earth. The only real danger to man is man himself.

How has he gained this supremacy? What is his present situation? What is his probable future? These are important questions. The first can be answered with some certainty. The second is hidden in the smog man himself created. The third is impossible to answer; but it is equally impossible not to speculate about it.

It has taken us a long time to become the masters of the earth. Collectively, we have acquired an enormous amount of experience. In a way, we can say that every cell in our body has taken part in the whole biological evolution, because every cell we consist of is in a sense billions of years old.

This is so because, when the cell divides, the same life processes continue in both parts of it. Hence, the life of the cells is an unbroken continuity from one generation to another. The cell has no mother; it is identical with its mother. All the cells in a person's body have come into existence as the result of successive divisions of the egg cell which was fertilized nine months before his birth. Every cell in the body lives a life which is a direct continuation of the life of the fertilized egg. With considerable justification, then, we can claim that every cell is identical with the egg cell (even if during its life it has changed its shape).

If we trace the lives of the cells back through time, generation to generation, we discover the same continuity of the life processes. These life processes constitute a direct bridge between some original cell in the dawn of time and those cells which constitute our bodies today. In this sense, every cell that is alive today has experienced the whole biological evolution.

Our oldest ancestry is identical with that of the amoeba Scheherazade. But the history of our cells has been much more complex. Our genetic line passes through one of the unicellular animals that chose not to be alone and joined with some friends to become a multicellular animal. In the course of successive generations in the sea, our ancestors developed into increasingly complicated animals. Finally a remarkable event occurred: the first lung fish crawled up on land. We participated in that event in the sense that the cells of which we consist are offsprings of an egg cell in that lung fish.

If we continue to follow our genetic line, we find that the lung fish mutated little by little in its effort to solve the problems of living on land. It learned to tolerate the dryness and to move about in it. The continents, which previously had been sterile deserts, were at that time just becoming covered by plants. The evolving terrestrial species began to find life on land endurable.

Many of our distant relatives thought that it was unnecessary and dangerous to leave the water, which is the primary element of life. They mutated to new species of fish, and they still swim around in different shapes and sizes. And, in reality, wasn't it foolish of our terrestrial predecessors to settle on land?

Certainly their start was difficult. They were compelled to change their manner of life. In the water, they had been weightless; on land they were compelled to grow legs to carry them, and they had to learn to cope with gravity. Because cells can live only in water, they had to bring the water with them, confined within a protective water-tight sac, namely the skin. It took a long time for them to become accustomed to the air, but once they did so, they became addicted to it. Now, none of us can retreat to the water, our original home, without being drowned.

Life on land proceeded according to the laws of biological evolution. The first terrestrial organisms mutated in different directions. Some developed into dinosaurs. Others remained small and insignificant, and lived in perpetual terror that the dinosaurs would destroy and exterminate them. However, it was the dinosaurs who disappeared. Various relatives of our ancestors acquired, as a result of mutations, qualities such as strength (and became predators) or swiftness (and became antelope, deer or the like); others lived in trees (as do the gibbon and the orang). Our own ancestors, however, mutated in such a way as to develop bigger and better brains than other animals. This made it possible for them to dominate the earth and to introduce a new epoch in the history of life. During this long history of evolution our predecessors mutated innumerable times, and in his genetic code, man now has stored all the knowledge he needs in order to survive. But it is not until the present stage in our development that we have been able to reconstruct our evolution.

BIOLOGICAL AND CULTURAL EVOLUTION

The whole biological evolution is a consequence of a very simple principle. A cell divides into two and each of the parts usually have the same properties as the original cell. This is repeated billions and billions of times. But on rare occasions, a mutation takes place, and a property is changed in some way. Those cells that are alive today constitute the living beings which, by accident, have a history of a long series of fortunate mutations that have allowed them to survive. (Our account is somewhat simplified. In reality, bisexual reproduction makes the conditions for survival more complicated.) All the other cells which have not mutated in a similarly fortunate way—an incredibly huge number—have been dead for thousands, or even millions, of years. From the point of view of life-from-the-outside, this accumulation is a simple result of probability. But if we consider it from the perspective of a life-from-the-inside, it should instill in us a fantastic wonder, the wonder that we are alive.

When an egg cell becomes fertilized, it begins to divide and form a complicated animal. This fascinating process is directed by the genes in the cells. They represent a sort of memory, which the species has acquired throughout its biological evolution. The experience in the art of surviving, which has been gained in the course of millions of years, has been stored in the code of the genes. This genetic code has a remarkable stability, which insures the reproduction of the species. It is true that it changes and "improves," but that is a very slow process.

Besides the memory of the genes, every individual has another memory—that mental faculty that we usually think of as "memory." This memory registers and stores our experiences, coding them in a complicated way, at least partly as electrochemical pathways in the synapses (the combining elements between the nerve fibers) of the brain. In this way we acquire and retain the knowledge of how to avoid a danger, how to find food, and other such experiences that are of decisive importance in the fight for survival. This synaptic memory has the advantage of working very rapidly so that an experience of today can be used tomorrow, whereas the genetic memory needs millions of years to change, to store new information. However, the synaptic memory dies with the individual. Thus it is only the genetic memory that the species conserves from one generation to the other.

The genus *Homo* was responsible for a very important invention. It invented the school. It found that man could use the synaptic memory as a memory of the species. Of all animals, he alone had the qualifications for this: a brain with enough capacity, and the ability to talk and hence transfer large quantities of information from one individual to another. In this way, the experience of every generation could be transferred to the next and used by it. Some knowledge—for example, sucking the breast of the mother, familiarizing oneself with things by putting them into the mouth to find out whether they are edible, fleeing when threatened, and copulating—is stored in the genetic memory, which constitutes some of the experiences of millions of generations. Other knowledge— how to make fire, how to fabricate tools, how to plow a field and domesticate animals, how to use weapons for defense or attack, and how to construct atomic bombs and travel in space—is information that mankind has acquired in a short time. Such knowledge is, in fact, the property of the species of man because from generation to generation it is transferred from one synaptic memory to the other. This inheritance of knowledge can be changed and increased from one generation to the next. Man need not wait for a thousand generations in order to allow the genes to incorporate new experiences. The synaptic memory is rapid. It is the school, then, which has supplemented the slow biological evolution by the much more rapid cultural evolution. It is the school which has made us the masters of the earth. In the first schools, the teachers were the parents, and the information they passed on to their children was what they knew themselves. Sometimes the children could learn also from others besides their parents. But when society became more specialized, particular areas of knowledge came to be possessed by different specialists. Artisans of different kinds—priests, gurus, teachers, and scientists—accumulated theoretical and philosophical knowledge in their particular areas of competence. The quantity of knowledge increased like an avalanche, and soon there was too much for any human brain to store. Then the brain was supplemented by libraries. Thus today it is surely reasonable to suggest that the key to our mastery of the earth is stored in our libraries.

FROM AN AGRICULTURAL TO A CYBERNETIC SOCIETY

The cultural evolution has already completed its first phase and begun its second one. During the first phase we lived in symbiosis with nature. We used it, enslaving the animals and exploiting the earth. Yet at the same time we were completely dependent on nature. We were part of it. But in the past one hundred years we have embarked upon a new phase, the technological or cybernetic era. This means that in the course of the past century we have broken our collaborative bonds with nature and produced a new milieu, which is more and more a product of ourselves. When most of us look out of a window, we see houses, streets, and cars—all man-made. Agriculture is no longer the basis of our economy. An advanced country must eventually cultivate industry, lest it risk economic impoverishment in the modern world. The cultivation of the earth and the domestication of animals yield only a small proportion of consumer items. Very few of the products in a supermarket come straight from a farm; most of them have been processed in factories. Farm products today become simply raw material for the alimentary industry. Attempts are made even to synthesize our food in a chemical way. We do not go out into the woods in order to cut trees when we feel cold; instead, we adjust the thermostat which regulates our oil heating. The house is a machine to live in, and if the electricity goes wrong, it becomes uninhabitable. The speed with which we can travel is much greater than that of a galloping horse. It is the speed of a car at seventy miles an hour, or of a jet at several hundred miles an hour. If we wish to talk with someone, we need not sit in the same room with him; he can be at his telephone on the other side of the globe.

The changes in our conception of the universe are equally important. We know that we do not live on a flat disk that is covered by the crystal spheres of the heavens, but on a tiny little grain in gigantic space which is measured in billions of light years. In particular, we know that our species was not created by a mighty God a few thousand years ago, but that it is the result of a biological evolution that began billions of years ago. Formerly, we believed that the soul with which we were endowed at the creation made us completely different from the animals. Not until now, when we are breaking the symbiosis with a biological milieu, have we begun to understand that we are part of it.

The drastic change from the agrarian to the cybernetic milieu is perhaps comparable to the change the first terrestrial organisms experienced when they left the water and began to live on land. Even that change entailed enormous difficulties before these early organisms became adapted to it. Many of their characteristics, which had been valuable for sea life, became useless or disadvantageous on land, and consequently completely new properties, the results of selective mutations, developed in order to save life. At present, man too is in the midst of a milieu crisis, which is similar to the earlier one in many respects.

However, the time scale for the agrarian-cybernetic crisis is different from that for the water-land crisis. The adaptation to land life took millions of years; the transition to the cybernetic milieu will be measured in decades, or at the most a few centuries. This means that the transition from water to land required thousands, or hundreds of thousands, of generations, whereas the transition from symbiosis with nature to the symbiosis with our own technology may take only a couple of generations. Many people living today were born in an agrarian surrounding, but by middle age they were suddenly precipitated into the cybernetic life. We have shifted from the steady trot of the genetic memory to the jet-age speed of synaptic memory.

There are other differences which are still more important. When those early lung fishes began to move from the sea to the land, they led a dual existence for a long time. They were able to live on land if this was more pleasant, but they could also return to the water. They had the option of retreating to the familiar and secure milieu if the land experiment did not succeed, since they were still equipped to live in the water as before. Not until their survival on land was insured did they lose their ability to return to the sea.

However, in the present transition, we are burning our bridges behind us. Even if we wanted to, we could not return to the pretechnical life. Rousseau's phrase, "Let us return to nature," has now become obsolete. We have already passed the point of no return. There is hardly any place left on earth where cars and airplanes are not producing noise and smoke, where the water and the air are not polluted, and where all natural resources are not being exploited. Of most crucial importance is the fact that there is no place where we can be safe from atom bombs and missiles.

If our technological symbiosis develops successfully, our life will become rich with possibilities we have hardly dreamt of. But if the cultural transformation goes awry, our species may become extinct as a result of atomic explosions and general poisoning. If the first lung fish found the solar radiation too hot or the wind too cold, it could crawl back into the water and protect itself. But where do we find protection from something that radiates more energy than a thousand suns, and from a death wind that is saturated with radioactive strontium? When biological evolution became mingled with the more rapid stream of cultural evolution, we began an adventure which may be either successful or catastrophic, and the outcome depends solely upon ourselves. The choice is our own. We may direct our cultural evolution or we can allow it to run amuck. Furthermore, we cannot ignore the possibility that our ecology is similar to that of the lake which destroys itself.

We have to choose between directing our fate and submitting to the "unavoidable."

THE EVENTS OF TODAY AND TOMORROW

In our technological era we are bombarded with so many demands on our attention that it is difficult indeed to distinguish between the essential and the unimportant. In fact, this distinction must always be dependent upon who the observer is. To me, certain events seem important; to you, perhaps they are not; and what may be interesting to both of us might be totally irrelevant to yet a third person living in another country, or in another culture remote from ours.

Suppose we read a newspaper and try to discover from the headlines what has really happened recently. In most papers, the latest football results are placed alongside a report on current space achievements. In addition, we find accounts of hijacking, kidnapping, assassination, student riots, racial conflicts, mass death by starvation in underdeveloped countries, and the division of the world into two conflicting camps—each arming in order to protect itself from the other.

Many of today's news items will be forgotten tomorrow; almost all of them, within a year. But this does not mean that they are without significance.

Historians look at, and draw conclusions from, the course of history in different ways. One historian will say that wars are won by generals, and another will insist that the men in the trenches deserve full credit. These clashing points of view are in fact reconstructions of historical events. Both of them illuminate important aspects of what actually "happened," but neither of them may have captured more than a fraction of the truth.

The sequence of history as we know it depends on a series of incidents and accidents that we might call "major" events: if a bullet speeding towards the heart of a general or a president were to change its course, then the course of history itself would almost certainly have been changed.

But "major" events are balanced by collections of "minor" events. Every new birth increases the population, every new factory increases the process of industrialization, and every new discovery increases man's collective knowledge. When we consider their cumulative effect, then, we see that overwhelming numbers of these minor events create a drift in the historical current that no leader, no matter how great, can swim against for long.

A historical incident is far too complicated and many-sided to be completely and truthfully described. Historians select—often quite subjectively—certain aspects of the incident and say, "this much happened and the rest we can forget about." Thus history becomes a reconstruction: an "image" of events the historian has created through selection and argumentation.

If it is difficult or impossible to reconstruct past events accurately, it is obviously even more impossible to predict what will happen in the future. But at the same time it is to a certain degree even more urgent to do so. Each of our actions is fraught with consequences for the future, and, for this reason, most of us act with the future very much in mind. We are like the donkey who walks because the carrot is in front of him. In fact, Pascal was probably right when he said that men live more in the future than in the present. Therefore we really cannot avoid being concerned about tomorrow. Perhaps this is why people flock to palm readers, astrologers, and crystal ball gazers.

It is interesting to ponder the degree to which we can predict the future on the basis of what we know about today's conditions. It is possible to predict some occurrences with certainty, and totally impossible to predict others. For instance, if an expedition departs for a distant place to make astronomical observations on an eclipse of the sun, its members can be sure that the eclipse will occur at the exact time that previous calculations predicted it would: it is a fact that many astronomical phenomena behave according to laws that are completely understood. However, those astronomers cannot be sure that they will be able to observe the eclipse. Perhaps the sky will be clouded at the critical moment. Who can predict the weather with absolute certainty? Perhaps the expedition's airplane will develop engine trouble. Perhaps a member of the team will become ill, and the group will have to return to home base. There seems to be an indeterminate number of possible obstacles that cannot be anticipated.

The problem of how to predict the future scientifically has not been broached until quite recently. But today a new science, often referred to as "futurology," is an attempt to do just that. This discipline has been received with considerable scepticism (any other reaction would be surprising), and many of its findings do indeed seem rather naive. But such a science is needed. Too often, our present predictions are based on the assumption that the same forces that determine events now will always determine them. Often we implicitly postulate that progress has attained its zenith in our time and so come to a halt; we assume that no new devices or processes will be invented, that no new significant ideas will be advanced.

Naturally, futurology is still at the stage of guesswork and general discussions about methodology. Man has a long way to go before he can systematically predict the benefits and the disasters that the future will bring—if such predictions are indeed even possible. Nevertheless, in our discussion of mankind's situation today, we cannot avoid an attempt to evaluate future possibilities, simply because man's actions today are always largely determined by what he expects of tomorrow. Though our ignorance of the future will permit us only a guess, it does seem that man may have ceased to stroll aimlessly on history's vast continent, and has at last found that narrow isthmus of time that joins the past to the misty land of the future: we may never reach the shores of this land, but perhaps from time to time the clouds will lift just enough to permit us glimpses of its silhouette.

III

Life and the Cosmos

THE COSMIC THREAT

In Chapter One, we suggested that there was an analogy between the biosphere, with man as its ruler, and the eco-system of a lake. We found that life in a lake could die from two causes: the outer conditions that produced the lake could change in such a way that the lake would be transformed into a desert and all life in it be extinguished; or life in the lake could destroy itself.

Now let us return to the biosphere, and begin by dis-cussing the first possible cause of extinction. Is life on earth likely to be extinguished by exterior forces? The cosmos, which once made the third planet inhabitable and produced life on it, may in the future make that planet uninhabitable and kill all of us. If we are to consider whether such an event is likely we must project ourselves into the future, using the time scale that we used in our discussion of the transition of our genetic line from water to land.

Let us therefore suppose that we succeed in avoiding the perpetration of genocide on our whole species. If human culture does indeed manage to evolve without suffering a final catastrophe, for how long a time can man persist as a cosmic phenomenon? The answer is that no end is in sight.

In the medieval ages, and even much later, many people feared that the earth would perish as the result of a cosmic catastrophe. A big comet might come and collide with the earth, they thought, or its tail might poison the atmosphere. But the risk of a collision is very slight, and the quantity of poisonous gas in the tail of the comet is negligible. If a collision should really occur, considerable damage would be caused, comparable perhaps to that effected by a couple of hydrogen bombs, but certainly in no way as disastrous as a full-scale atomic war. Man is not really threatened by comets. In fact, the threat of comets has faded in the light of new threats which we ourselves have created.

The danger presented by asteroids is equally negligible. A few years ago there was a rumor that the asteroid Icarus was going to collide with the earth; when its orbit was calculated more accurately, it appeared that (measured in cosmic units of distance) it would pass quite close to the earth, but still at a completely safe distance. It came, it was observed by the astronomers, and it disappeared again. It will return many times to the neighborhood of the earth, but the chance of a collision is less than one in a million. If there really should be a collision, great damage would occur, but the existence of mankind would not be in danger.

Meteors, which we observe every clear night, consist of small grains of sand which evaporate high up in the atmosphere. Only seldom does it happen that rather big stones strike the earth. Huge meteorites have produced large craters, such as Meteor Crater in Arizona. If a comet or an asteroid should collide with the earth, it would look like a gigantic meteorite. In 1908 a large meteorite fell in Siberia and destroyed an area many miles in diameter. But such catastrophes are local.

There have been best-sellers which fabricate the myth that Venus or other planets were once very close to the earth, causing catastrophes, and that a similar catastrophe might occur again. Such a conjecture is nonsense. We have determined the orbits of the planets—what they were in the past and what they will be in the future—with sufficient accuracy to exclude the possibility of such events having happened, or of their happening in the future.

CAN WE TRUST THE SUN?

But are we threatened by the sun, the great creator of life, on which we are totally dependent? It has been claimed that the sun can explode and burn the whole earth to a cinder, or that it may burn out, so that all life will freeze to death. Are these prophecies correct?

We have studied stars of many different kinds. Certain kinds can explode and thereby become what we call a "nova," a star whose emitted light increases greatly for a short period of time and then returns to normal magnitude. Other types of stars undergo much more violent explosions, and these are called "supernovae." The sun, however, belongs to one of the most stable categories of stars we know. With a very high degree of certainty we can predict that it will not explode or be extinguished for the next few billion years. We cannot, however, be assured that its light intensity will not change. In the past the earth has seen long periods during which the climate in large parts of the world was much colder than today. We do not know with certainty what caused these periods of extended glaciation, the so-called ice ages. It is possible that changes in solar radiation caused them, but some scientists believe that the ice ages were a meteorological phenomenon. Perhaps the atmospheric circulation may have changed. Alternatively, the radiating conditions may have altered and caused an imbalance in the so-called greenhouse effect: an effect by which the atmosphere keeps the earth warm in much the same way as the glass keeps the interior of a greenhouse warm. There may very well be new ice ages in the future. If the northernmost countries should be covered by a layer of ice a mile thick, this would obviously be very unpleasant for the people living there. But it would not constitute disaster for the whole of mankind, not even for those who live in such countries. A new ice age would certainly not come suddenly. It would probably be preceded by one or more centuries of gradual freezing, and thus there would be ample time to organize an emigration. Also, man may well be in a position to change the climate. Our increasing technological knowledge will probably make this possible in a not too distant future.

Consequently, it is safe to conclude that we can trust the sun for billions of years. It may give us small troubles, but it will not threaten the existence of man. The sun god protects us and is worthy of our worship. What may happen after several billions of years have elapsed is of no concern for us today.

Or is it? Perhaps the destiny of mankind in the very distant future is not altogether irrelevant to us. Most of us who lack the consolation of believing in personal immortality feel it is essential that something of us will survive our death: if we have children, we think of them as a continuation of ourselves; if we do not, we hope that something that we have done in life will survive us. We believe that our thoughts and actions have implicitly contributed to history, to the evolution of mankind, and that we are responsible for the collective karma of man. But if the final end of mankind is to be total annihilation, and the sterilization of the earth itself, what then is left of our faint glimmer of hope in man's immortality? To those of us who have this philosophy, it does not really matter whether the end of mankind comes within a few thousand or a few billion years: in both cases we are thinking in terms of very long time spans, impossible to comprehend very realistically. What does matter is whether the final result of all the work done by mankind in the course of innumerable generations has ultimate significance. Let us therefore consider the final destiny of life.

Even if we can trust the sun for a few billion years, it is inevitable that it will finally burn out. It gets its energy from the fusion reactor in its center, which converts hydrogen into helium. The sun has enormous quantities of hydrogen (the major part of it consists of this element) and it consumes it at a steady but not too rapid rate. There is enough to guarantee continuation of the solar energy outflow for several billion years, but finally it will run out. What could we do then?

Until the beginning of this century the answer was irrefutable: we would freeze to death. Without the sun the temperature on earth would drop to absolute zero. Space would take us in its icy grip. Mankind, dependent upon an agrarian culture, could not avoid extinction.

But now we have made the transition, or are in the process of making it, into the cybernetic era. We have begun to understand the cosmic forces so well that we can exploit them. We are no longer a ping-pong ball to be batted around by them; instead we are learning to direct them technologically. We will be able to do so, perhaps not immediately but within a span of centuries; that is, within a very short time compared to the length of time the sun will condescend to grant us.

In science fiction literature nothing is viewed as impossible. Even if our planet should become uninhabitable, writers of science fiction offer various solutions: we may simply move to another planet; or we may settle down in space; or, if the sun burns out, we may construct an artificial sun. How much truth is there in such ideas? Let us see what the theoretical possibilities might be, even if it would take a long time to make them a reality.

INHABITABLE PLANETS

We can already travel to the moon, and it will not be long before we are able to visit Mars and Venus, though none of these celestial bodies are presently inhabitable. But if we made full use of our technological knowledge, it is possible that we could make them suitable for man.

When life originated on earth the atmosphere consisted of nitrogen and carbon dioxide, and it is probable that no oxygen existed at all. Therefore, in the beginning life had to be anaerobic, since only anaerobes, the first microorganisms, were able to live without oxygen. They absorbed carbon dioxide from the air, retained the carbon in order to build up their cells, and exhaled the oxygen. The plants of today live by a somewhat similar process. Carbon has, throughout the geological eras, been deposited in the immense layers of coal, and perhaps also in the oil, that exist in geological formations. The oxygen released through the life process of the plants, and through related processes, has accumulated in the course of long periods of time, until it now constitutes one-fifth of the air. Not until this process of oxygen release had continued for a long time and a sufficient amount of oxygen had accumulated, was it possible for animal life to develop, since animals are dependent upon the oxygen which the plants breathe out. It is the microorganisms and the plants which have made the earth inhabitable for biological creatures like ourselves.

It is reasonable to assume that our cybernetic culture may some day be capable of achieving the same thing microorganisms accomplished long ago. Human activity on earth has already reached such proportions that it continually changes the conditions of life on our planet. Exhaust gases from automobiles poison the air, not only in big cities, but throughout whole states, as in California. Not only are seas and rivers polluted, but it appears that even the Pacific Ocean is too small to serve as a garbage can for, say, radioactive wastes. There are enough hydrogen bombs stored in various parts of the world to poison the whole terrestrial atmosphere if exploded during a nuclear war. Since the beginning of the industrial revolution we have burned so much coal and oil that the amount of carbon dioxide in the air has increased so greatly that it may be one reason why climatic changes, for instance, have occurred.

Our activity on earth is perhaps eventually going to make it uninhabitable. We, of course, did not intend this, but it appears that the forces we are now releasing are of almost "cosmic magnitude." Our growing technology is allowing us to change the very character of our planet. In the near future it will probably enable us to change the character of our neighboring planets as well. It is possible that we may again need the help of our friends, the microorganisms. After all, it was they who started life on earth. They came into existence on a sterile planet and made it inhabitable. They have not been able to start the same process on Venus, perhaps because the planet is too warm.

It is possible that if we help them with an initial change in climate, if we direct the process in a suitable way, we may in fact help to create a new inhabitable planet with an atmosphere that will sustain life. Mars, too, can be changed. At present it is too cold, and the oxygen content of its atmosphere is too low, to support any kind of animal life. Human ingenuity may find a way to make it inhabitable.

But what do we really mean by an inhabitable planet? We are rapidly advancing in the field of biological experiments. We can transplant kidneys and hearts and soon we will start to change ourselves in more drastic ways. If we find it difficult to change our neighbor planets in space, perhaps we may change our own species in such a way that we will indeed be able to exist on other planets, or on a decaying earth.

LIFE IN SPACE

We can also settle down in space. Some astronauts have already managed to subsist there for periods of several weeks. Space ships will soon be more comfortable, and what the moving picture *2001* has depicted will be reality in the near future. Permanent scientific observatories, to be manned by a crew which is relieved at intervals, are already being constructed for location in space. The whole of our planetary system will soon be investigated from such observatories as these movable bases.

Hence we can look forward to life in space. Here, "we" does not necessarily refer to the whole of humanity, but rather a very small fraction of it. Life in space will probably differ less from our accustomed existence today than does today's existence from life in the agrarian era. If we could go back into the nineteenth century to explain to a farmer how one is able to live in the cabin of a jet plane, it would be difficult for him to understand and still more difficult to believe. There was no one at that time, with the exception of Jules Verne and Edgar Allen Poe, who could even "leap" imaginatively from the agrarian to the cybernetic era—yet it took less than one hundred years for mankind to make the change. Similarly, all the display of imagination in science-fiction literature notwithstanding, it is doubtful that we can successfully imagine what space life will be like in the not too distant future.

Before long we will probably become accustomed to living elsewhere within our solar system and its close environment. But once the solar system has been colonized, the next step will be an enormous one. Light requires four years to reach the next star, and a space ship with its present velocity will take one hundred thousand years. Our next neighbor, then, is really far away. But we must not forget that technology advances at an increasingly rapid rate. Scientists are already discussing ways in which the velocity of space ships might be increased, and the cost of sending a space ship to the next star has been estimated.

But in reality, why should we send an expedition to a distant star? Why should we go there? The same silly question was asked when the first moon-ship was launched. Columbus had to answer the same question when the *Santa Maria* weighed anchor. This recurring query goes back to still an earlier time: when the first lung fish crawled up out of the water and started life on land, his neighbors were undoubtedly shocked, and asked, "Why should he go there?" Indeed, if someone had told the lung fish, while he was lying in the mud at the shore, of golden waves of grain and whispering woods, then even this remarkable pioneer might have shaken his little head and muttered, "Incredible, incredible, and in any case there is no point in going there if there is no mud."

A NEW SUN

What the technicians call our energy consumption (in coal, oil, water power, etc.) is only a small fraction of one percent of the energy which we really need for a comfortable life. The major part of the energy that we ultimately consume heats the ground, the air, and the sea, and makes the farmlands ripen and the woods grow. The sun is the source of all this energy. The total energy consumption by technology is an almost negligible fraction of the energy the sun delivers to the earth.

When the nights became too cold in those tropical regions where man made his first home, he found that he could warm himself by burning wood. The discovery that he could control fire made it possible for him to settle in cold regions in which he could not depend on the warmth from the sun. His need for energy was met by going out in the woods and cutting trees, and then releasing the solar energy that the trees had absorbed during the past decades. But the cybernetic culture is ravenous.

It demanded new sources of energy, and found them in coal and oil, thus beginning the depletion of the energy accumulated during geological times.

Let us now return to our original question, which admittedly is purely academic. What would happen if the sun suddenly went out? If this unpleasant event had occurred in the nineteenth century or earlier, when agrarian culture was predominant, all life on earth would have expired in a very short time. In a few weeks or months everything would have been frozen. Piles of coal or wood could have saved perhaps a few people for a year or two.

Today, our situation would be slightly better. Fossil fuels and atomic energy might keep small groups alive for a somewhat longer time. But soon even these energy sources would be used up.

Our situation will change drastically once we are able to release hydrogen power in a fusion reactor. Then we will have access to the enormous quantities of energy that can be extracted from hydrogen, the most abundant element in the universe. At first, we will be able to use only the heavy hydrogen, but later we will also be able to make use of the light, more abundant isotope. When we can construct a fusion reactor, we will be able to reproduce the nuclear burning that now takes place only in the central parts of the stars. In this way we will create artificial "miniature suns," energy sources which will make us independent of the sun. We can exist very well with the aid of energy generators that are small by cosmic measure.

Once we have discovered how to manufacture such a fusion reactor, then, we will indeed become the masters of the cosmos. We will be able to travel freely—within the limits set by the enormous distances between stars—and we will be able to liberate as much energy as we consider necessary. Under such conditions, if our sun should die, we could easily survive. But we would naturally miss her and bewail her death.

THE REAL THREAT

As far as one can see there is no real danger threatening us from the cosmos. Our cybernetic culture is well on the way to making us masters of space. Our species is potentially immortal.

Are there really no disasters threatening us? Actually, we are very seriously threatened, but the only real danger to man is man himself. We may annihilate ourselves.

IV

Man's New Environment

THE NEW SITUATION

We have found that the cosmos does not threaten us seriously. But what about the danger from within? Will man kill himself by his own actions, just as life extinguished itself in the aging lake? If we are to attempt an answer to this question, we must completely change our way of looking on man. We must consider him not as an astrophysical phenomenon, but rather as a biological being, a part of the ecosystem of the biosphere; still more important, we must consider him as a human being—not a unique individual, but a member of a society which he has formed and which is forming him. The focus of our studies will have to be shifted from the natural sciences to sociology and related fields.

We realize that man's potential for domination of space has demonstrated how powerful our cybernetic technology is. We have overcome gravity, which until now has kept us fettered to the earth's surface. Soon we will be able to release and exploit hydrogen energy, thereby obtaining the power reservoirs that we need in order to broaden our domain in space, using the earth as launching pad. If a culture can master forces of such magnitude, we might reasonably expect that culture to have the ability to organize a peaceful and progressive life on the planet on which it originates. But what do we find if we peer down at earth from the perspective of space? We see a general dissatisfaction with the present state of affairs and a fear of what the future will bring us. We see incredible material abundance in some parts of the world, but great misery and starvation in others. We see, above all, the tremendous fear of atomic bombs, missiles, bacteriological warfare, and so on. As we analyze man's present situation, we might ask ourselves if our problems are not in some respects like those men faced two hundred years ago. The people of that time were also afraid that a catastrophe was approaching, yet had no idea of how to avoid it; there seemed to be little they could do but await its coming with the far from consoling thought, "après nous le déluge."

Although our technology has shifted us from an agrarian culture to a cybernetic one within a very short time, the political structure of our globe remains fundamentally unchanged. The new forces which science and technology have unleashed are used for the same old political purposes—to repress, to kill, to gain power. The politicians seem hardly to realize that a new era has dawned.

FIVE AGES AND TWO EXPLOSIONS

During the past hundred years the industrial revolution in technologically advanced countries has drastically changed not only the outer life of man, but the inner life as well. In other words, man's view of himself has altered. New methods of production have abolished general poverty and the shortage of necessities, and have given large numbers of men an abundance of goods. Moreover, human working conditions, living conditions, and the relationships between men have also drastically changed.

The new energy sources have given us a global system of communication and we measure the traveling time between continents in hours instead of in days, months, or years. But political conflict and warfare have now assumed global proportions, and people everywhere feel the existence of humanity in danger whenever political conflicts threaten the outbreak of yet another war. This world-wide danger necessarily causes us to begin to feel, think, and plan on a global scale and to recognize that global cooperation is a necessary condition for our continued existence.

But the changes we are witnessing are so many and so fundamental that it is very difficult indeed to analyze them.

There are a number of new concepts, some of which

are commonly expressed by catch words or slogans which summarize certain characteristic features of our time. Let's examine a few of these. The term "space age," for example, refers to man's attempts to dominate space, which we discussed in the preceding section. The expression "jet age" denotes the constantly increasing speed of the transportation systems which have diminished the distances on the earth. We might say that the term "atomic age" alludes to everything that has been achieved through the application of atomic physics. But the most calamitous consequence of the atomic age has been the atomic bomb, which necessarily lends a special character to the word. To many people the atomic age means the period when mankind discovered the means to annihilate itself. The advances of chemistry have made it possible to synthesize a number of new substances, thus allowing us to replace products which nature previously provided for us. Since we lack a generally accepted term for this activity, we might designate it, in a general sense, by the phrase "synthetic age." A fifth "age" might be called the "computer age," suggesting the computer's immense implications for a changing society.

The sixth characteristic of our changing environment is the increase in the numbers of people throughout the world—an increase that is causing our planet to be, if not already overpopulated, well on the way to becoming so. This growth is called the "population explosion."

The terms that we have just listed refer to a number of drastic changes in man's environment. What, then, is the reaction of the individual to all of this? The reaction which is most important for our future may properly be

termed the "expectation explosion": the increasing demand of each individual to receive his share of the benefits of science and technology—a frustrated demand in
many cases.

In the remaining sections in this chapter, we shall examine in more detail these seven terms and their significance for us today.

THE SPACE AGE

It is in this age that the positive achievements most clearly
outweigh the negative ones. Certainly the technological
heritage of the space age may be traced back to the construction of missiles (awesome and terrifying weapons)
during the Second World War. These missiles did not decide the outcome of the war, but they now help to make
the world's political situation very unstable. But the development of missile technology from military missiles to
space ships has not been made exclusively for military
reasons, because the Moon, Venus, and Mars have very
little value as military bases.

As we have suggested in preceding pages, the construction of space ships makes the colonization and exploration
of space and of the celestial bodies in our solar system
possible in the same way that the transition from life in
water to life on land opened up new possibilities so long
ago. Should we find space life agreeable and interesting,
we could begin to acclimatize ourselves to space and
build self-supporting colonies there. If we let our imagi-
nations run free, we might even hope that our advances in

space technology would permit a small proportion of our species to be rescued, if we did indeed precipitate a catastrophe here on earth.

THE JET AGE

The jet age, symbolizing more rapid modes of transportation and communication, is Janus-faced. On the one face, we see that those who travel to foreign countries appreciate rapid transportation and it is generally thought that better means of mass communication ought to increase understanding between different peoples. But on the other, we see the possibility that the more rapid means of transport are increasing the frictions between different nations and thereby increasing the risk of war.

In earlier times, a war between, say, the United States and Russia would not have been a very important event. At most, some minor forces would be sent to the periphery of the adversary's country. The journey from Washington to Moscow was very difficult and time-consuming, and it would have been impossible to inflict vital destruction upon the enemy.

But the construction of heavy bombers changed this situation. When the jet age began, each of the two capitals was brought within several hours' reach of bombers from the other country, and the introduction of missiles has shortened the time still more. Submarines armed with guided missiles are now permanently on watch, and if one country's supreme commander were to issue the order, a large portion of the enemy population would be killed within an hour.

The accomplishments of the jet age, then, have made the earth shrink. They have made it easy for us to meet our friends, even if they live on the other side of the planet. They have also made it impossible for us to escape our enemies, even if these too live on the planet's other side.

THE ATOMIC AGE

Our acquisition of knowledge about atomic structure has been of decisive importance to the whole technological era. Atoms consist of two parts: the central atomic nucleus and the surrounding electronic shells. The structure of the shell was the first to be elucidated. The application of this knowledge has had enormous technological consequences, because it has been the foundation for the production of all the new materials which have made the space, jet, synthetic, and computer ages possible.

The development of nuclear physics came later. From a technological point of view, the most important achievement in nuclear physics has been the release of the energy which is stored in some nuclei, most notably the uranium and plutonium nuclei. This energy can be released in an explosion, such as that of an atomic bomb, or in a continuous process which takes place in an atomic reactor. In both cases the energy release results from the fission of the nuclei.

The technological application of the uranium fission has made it possible to annihilate mankind. At present it is even a reasonably simple procedure. In both the United States and the Soviet Union there are large stores of atomic bombs. In each country the stockpile is more than sufficient to kill the whole of mankind. The technical term for this overabundance of nuclear weapons is "over-kill." In spite of the existing quantities, new atomic bombs and more sophisticated devices for their delivery are being produced. Only as long as we are confident that the stock piles are really under control and that those who control them do not intend to use them, can we sleep quietly.

A number of unfortunate coincidences have made the atomic bomb possible. The properties of a few atomic nuclei are such that they can produce a chain reaction, provided that they are aided by neutrons that move so rapidly that the reaction becomes explosive. Another un-fortunate coincidence is that the reaction can take place in even a small quantity of uranium or plutonium, so that the technological difficulties in making atomic bombs are not prohibitive. Furthermore, the supply of uranium on earth is quite abundant.

The peaceful use of atomic energy has given us a new source of energy, which is being used increasingly. We must not forget, however, that the production of atomic energy is necessarily accompanied by the production of large quantities of extremely poisonous radioactive sub-stances. From a theoretical point of view it is possible to bring these waste products under control by storing them in such a way that the risk of poisoning life on earth be-comes minimal. However, this storage would require an extremely rigorous international control organization. It is highly doubtful whether it is really possible to over-come the difficulties of controlling the radioactive waste.

The atomic age deserves to be called "mankind's night-
mare."

THE SYNTHETIC AGE

Synthetic products, such as plastics, are rapidly replacing
more and more of the material which nature once provided for us. Instead of building a boat of material which
trees have synthesized from sunlight, air, and water, we
now cast the hull out of a petrochemical product. Other
synthetics are replacing wool, cotton, and silk.

The synthetic age may be said to symbolize the way in
which man is ceasing his symbiosis with nature, and his
exploitation of plants and animals, and is instead beginning a collaboration with his own technology. This age
offers many advantages, one being the everyday pleasure
which new and better products give us. We have good
reason to thank our chemists if they are truly able to surpass the silk worm and the cotton plant as manufacturers
of the materials for our clothes, and the trees as deliverers
of the substance for our kitchen tables.

But the synthetic age, as defined in our general sense,
also has many serious disadvantages. New technical processes give us useful products, to be sure, but they also
yield as by-products a number of poisonous and dangerous substances. Even the useful products remain useful
only for a short period, after which they are transmuted
into waste which we want to dispose of.

The great quantity of wastes, the ultimate products of
our industry, is causing an increasingly serious problem

for us, partly because much of it is not recycled by nature. Thus, our technology, which is helping to make us independent of nature, is also, unfortunately, causing the devastation and pollution of nature.

THE COMPUTER AGE

The age that is most difficult to analyze is the computer age. The term, used in a general sense, can be said to symbolize the whole cybernetic era. The first important function of computers was solving mathematical problems in scientific work. Also their military application was recognized soon after their invention. In the event of a technological or cybernetic war, an increasing proportion of the tactical operations will be taken over by computers. The guiding of missiles and torpedoes toward enemy targets and even the direct shooting will all be entrusted to computers because these react so much more rapidly than human beings. However, initial attempts to leave the strategy of a war to the computers do not seem to have been very successful. Nevertheless, in government and business administration the computers have proven indispensable because of the rapidity with which they work and their enormous capacity for storing information.

The computer age also has many unpleasant aspects. In many countries information about every citizen has already been stored in the memory core of some computer. The computer can remember, when called upon, not only the person's name and birthdate, and those of his parents, spouse, and children, but also much about a number of other aspects of his life: his state of health, his education, his qualifications, his ability for different kinds of work, his adaptability, and perhaps also his political views. Each national administration, of course, tends to increase its information about every citizen, and the present trend is quite clear: in a modern nation, more information about every citizen will be compiled and stored each year. This information is often of great value to the administration, but there is a fearsome risk that it will make political repression more easy. The computer technique makes it possible to measure and weigh us all and register every step we take in life. Hence, it makes possible a centralization of power that was not technically feasible earlier.

The computer age is very young. Every new "generation" of computers will be technically superior to the earlier generation. The experiments which are now being carried out in the leading laboratories suggest fantastic new possibilities. We may have started a completely new type of evolution, which has some similiarity to biological evolution, but is based on semiconductors instead of protoplasm.

Such speculations have prompted discussions on whether or not the computers may some day attain such a degree of complexity that they will in fact acquire "souls." Is it possible that they might possibly become the masters of the men? Is the computer perhaps the Übermensch, which Nietszche predicted? One may perhaps dismiss such fantasies as completely unreasonable. No one has ever discovered any soul in a computer. But we must remember that the computers we now have are only the third or the fourth generation of computers. What will happen when we have reached the one hundredth generation? Will the evolution still be governed by our synaptic memory? Or will the transistors in computers be in charge of the evolution? Will the transistor memory be actually an advanced third stage in the development of memory, following the genetic and synaptic memories?

THE POPULATION EXPLOSION

The sixth dramatic change that we are presently witnessing is the steady growth in the world's population, commonly known as the population explosion. The basic facts about this on-going process are too well known to require repetition here. In brief, the present sharp increase in the earth's population is partly a result of the advances in medicine that have made it possible to cure illness and lower the risk of death at an early age. This progress has given man more safety and dignity. But because the long range effects have been neglected, a significant proportion of the gain has been negated. Infant mortality has decreased; the lives of many children are saved, but they are saved only to become lives of misery. For what is a man profited, if he shall gain freedom from disease, and lose his life by starvation?

It is easy to say what ought to have been done in the past and what should be done now. Shortly after medical science had made it possible to reduce mortality significantly, it also made it possible to reduce the number of births without any serious consequences. It is obvious that one of the most important problems we face today is how to optimize the population of the earth. We shall discuss this problem more closely in Chapter Nine.

THE EXPECTATION EXPLOSION

Increasing numbers of people are discovering that the forces which science and technology have released can liberate them from misery and toil. The harvest of the fields has multiplied; industry mass-produces consumer goods which have previously been reserved for only a wealthy few or have belonged to the realm of dreams. What is the pilgrim's horn of plenty compared to a supermarket? And who does not prefer a jet plane to a flying carpet? Add to these the fruits of modern political liberty. In both democracies and dictatorships, those in power claim that every move is toward the realization of the avowed goals of the welfare, freedom, and happiness of the citizens. Is it any wonder that everyone looks forward to a prosperous future; that man's expectations are exploding?

But where has the abundance, the liberty, and the happiness gone? It is true that large population groups, both in the West and the East, enjoy a material prosperity that was previously unknown. But it is also true that the number of hungry and destitute people is greater now than ever before.

It is true that large groups of people have more freedom of speech and more freedom to do what they want and go where they want than earlier. But the suppression of freedom has also increased, and it seems possible that repressive tendencies are growing even within those nations that have democratic forms of government at present.

It is therefore easy to understand that the increased expectations necessarily lead to frustration and alienation within contemporary societies. One current manifestation of this dissatisfaction is a series of protests and revolts. It is possible that sooner or later these demonstrations will succeed in forcing the reorganization which is essential in order to adapt ourselves to the transition from agrarian to cybernetic culture. In fact, the new attitude engendered by the expectation explosion may turn out to be humanity's greatest asset in its present condition.

V

Smoke Screens

DELIVER US FROM EVIL

Every attempt to clarify and seek solutions to the global problems in human collaboration—to identify the dangers that threaten us and determine how to avoid them—is made more difficult, or even prevented, by various smoke screens. Like morning mists, some of these screens remain from a night when every event was attributed to supernatural or magical forces. Today, they continue to carry fatal consequences because they prevent us from seeing clearly. They allow us to ignore the unpleasant tasks which a clear-headed vision of modern life would force us to undertake. We hide behind the smoke and fail to act responsibly.

Still other smoke screens billow out from the chimneys of political and economic propaganda. These screens are calculated to shroud from our view political and economic measures which would be far from attractive were they to be seen in broad daylight. We can state without exaggeration that the modern atmosphere is indeed polluted, and that much of this pollution is intellectual rather than chemical. Whenever moral, political, and social views conceal reality from us, we are justified in suspecting that the future of mankind is at stake. Perhaps the most dangerous smoke screen of all is the defeatist-escapist attitude, an attitude maintained by those who run from responsibility by deciding in advance that the only solution is that there is no solution. According to this view, the world, and especially man, is inherently evil. From this premise it follows that it is not worthwhile to try to improve the world. Consequently, people who subscribe to this philosophy propose something like the following: let us therefore leave the world, enter a monastery or laboratory or business office, and concentrate on our daily work. Let us enjoy the happiness of home life without bothering about what is happening in the world. War is terrible, but there is nothing we can do to prevent that. It is like a natural catastrophe. A flash of lightning strikes our house and it burns, or a flood occurs and inundates our village; similarly, someday someone pushes the fatal H-bomb button. All these are catastrophes against which there is no protection.

Some escapistic views are symptoms of social irresponsibility. Others are remnants from an earlier period in which man was indeed a helpless victim of various kinds of natural catastrophes. Today we are able to protect our house from a flash of lightning with the aid of a lightning rod. Irrigation helps prevent drought, and dams protect us from inundation. The man-made catastrophe, war, is caused by the decisions of certain leaders, and it is not at all impossible for us to find means to protect ourselves against it too.

But if this morning mist is lifted, many will find the light distressing, since it will be more difficult for them to justify their moral indifference or passivity.

According to another view, only part of the world, though admittedly the bigger part, is evil. The land where we ourselves live, and the group to which we belong embody good and reason in a world in which the rest is evil and mad. Therefore it is necessary to support our leaders, strengthen our ideological attitudes and military preparations, and expose the devilish plans of our enemies. Such views are often propagated by leaders who find it to their advantage to strengthen their position by heaping coals on the fire of nationalistic, religious, or ideological fanaticism. Such feelings are also popular in the military-industrial complexes in many countries because they enhance the power and reputation of these bodies. Moreover, it is easy to produce good arguments for such views. In a capitalist country, for example, a list of all the evil things that the communists have done is considered sufficient evidence of the superiority of that country, and in a communist country it is equally easy to present an altogether true list of the terrifying misdeeds of the capitalists. But if one offers or accepts such an account as proof of the supremacy of one's own country or group, he simultaneously erects a smoke screen to conceal the flaws and guilt of that country or group.

A third view is one of naive confidence. To be sure, it is deplorable that there is war in some distant countries, but this is not of significant import to us. Our politicians are certainly much too prudent, and much too sensible, to engage in a new world war. It is true that there are large stockpiles of dangerous weapons, but of course they are under satisfactory control. Why should we listen to Doomsday prophets when in reality the world is not in bad shape at all. My family and I are better off than we have ever been and if everybody minds his own business and works for the progress of mankind, there is certainly no danger. Moreover, we are protected by Providence.

This is exactly what the amoeba Scherherazade said just before her death.

AGGRESSION AS A BIOLOGICAL INHERITANCE

It has often been claimed that war is an inevitable consequence of an aggressive instinct that is innate in man, as in many other animals. According to this view, war is impossible to abolish because this aggressive instinct in humans is so deeply rooted.

New intelligence on these views has been provided by recent investigations of aggression in different animals, investigations which have invalidated a number of common and dangerous misunderstandings. It has been pointed out that it is essential to make a distinction between killing and aggression. Every carnivorous animal—and most people do eat meat—must kill in order to live. Such killing is not, however, necessarily tied to aggression. For example, a fox who gets his food by killing a hare is not necessarily more aggressive than the hare who gets his by eating vegetables. Likewise, a farmer who sells his favorite cow to the slaughterhouse need not be aggressive. The butcher who slaughters the cow need not be more aggressive than any other worker. And the people who gather on a Sunday afternoon to feast on porterhouse steak may be as peaceful as a gathering of vegetarians.

FIGHTING AND WARMAKING

Another important distinction is that between fighting and warmaking. A fight occurs between two animals or two men, or perhaps between a number of them. Aggression can frequently be the cause of a fight, or it can be manifested by one or more parties during the fight. An aggressive attitude intimidates the adversary and may give the aggressor a certain advantage. The aggression, in other words, makes him psychologically and physically more adapted for fighting. Personal combat was often of decisive importance in the turbulent battles of earlier times, and it is still important in guerilla fighting. But the mechanization of warfare has diminished the importance of such fighting, and replaced at least some of it by the combat between mechanical weapons controlled from a far distance. The soldiers who handle these weapons are engaged in an activity that is not unlike operating the machines in a factory.

In order to make a soldier brave in hand-to-hand combat, it was essential to develop in him a sharp antagonism toward the enemy. But a soldier who controls a long-distance weapon ought to be a clever engineer, and he will work most efficiently if he is calm and not aggressive. Hence, in modern fighting, aggression is not as necessary as it was earlier. Individual combat is important and is often the decisive factor in a war, but warmaking in general is mainly an activity of a different kind. It is a large-scale social activity that requires the organization of a great number of people who can collaborate in an efficient way. A soldier in any army is in actual combat for only a small part of the war, and he need not even be prepared for fighting much of the time. For the short span during which he is actually confronted with the enemy it may be advantageous for him to be aggressive; but for the duration of the much longer period of time when he is with his company, comrades, and officers, he should display as little aggression as possible.

IS MAN A PREDATOR?

It has been claimed that war is an inevitable consequence of man's supposed predatory nature. Hence a great military commander is often called a lion. But such a comparison is preposterous. In a war it is essential that the soldiers be trained to obey a command and act according to a special pattern. Typically predatory animals, such as cats, lions, and tigers are very difficult to domesticate and train in a systematic way, and it is impossible to organize them into an army. A predatory animal is even less able to lead an army. A military commander who acted like a lion would be completely incompetent as a war leader, since the task demands intelligence, foresight, and calmness under pressure. But it is a fact that men may be domesticated and trained as if they were dogs, horses, monkeys, or dolphins. It is this fact which makes it possible for modern nations to initiate and to carry on wars.

WAR AS A SOCIAL ENTERPRISE

War is actually an organized social enterprise. In reality there is very little difference between compelling a group of people to build pyramids, the Great Wall of China, Roman aqueducts, or the Panama Canal, and inducing them to live in trenches. All these activities were very arduous and a large proportion of the people taking part in them died. Only drastic measures could force people to engage in such work. Existence in the trenches and construction of the Great Wall of China can be classified as military enterprises, whereas the others we mentioned were of a civilian nature. Though it is possible that aggression was an important reason why Vikings, pirates, and mercenaries went out on military expeditions, the lust for adventure and the chance to get rich quick were more important. Modern war does not begin because soldiers are anxious to go out and fight, though no doubt some officers really long to practice their profession and gain rapid promotions. But in a war it is not necessary that even the highest commanders be aggressive. In fact, it often happens that their feelings toward their adversaries have a touch of comradeship. In earlier times, commanding officers were able to socialize together with conviviality at an evening banquet and then command their soldiers to kill each other the following day. Warfare is not so "civilized" today. But even in modern times a captured general is often treated very graciously by his victorious adversary. They can sit together calmly, discussing the series of moves which led to the decisive outcome of the game of chess they have just completed. If they have any aggressive impulses they may release them by delivering sharp rebukes to their own orderlies more readily than by giving vent to antagonistic feelings toward those who are adversaries but social equals.

Neither is it necessary that the engineers who design weapons be aggressive. The men who designed the atom bomb were no more aggressive than other men; rather, they were scientists engaged in the work of completing a scientific task. An engineer working for a chemical plant may develop napalm one day, and a fertilizer the next. What is the difference? Business is business. The businessmen who speculate in wars and profit from them, and are in part the cause of them, are not necessarily more aggressive than other businessmen.

In every country there are always conflicts between individuals or groups which entail psychological or physical violence, these conflicts ranging from fights between boys to battles between business enterprises and political parties. These types of activities have little in common with warfare. Although considerable aggression can be discharged by large numbers of people, the harm caused by such encounters is negligible compared to the destruction caused by a war. If it is indeed biologically necessary to give man an outlet for his aggression, this could be arranged in many ways without war.

Hence, there is no indication at all that man's aggressive nature must lead to war. History shows us that there are many countries which have not participated in a war for several generations. It is possible for people in such countries to be born when the country is at peace, live their whole life in a peaceful country, and die before a war has started. It is evident that if the alleged human aggressiveness exists it can find an outlet other than warfare.

Among those countries which have enjoyed long periods of peace, Japan probably holds the record because it has enjoyed two long spells without any armed conflicts with a foreign country. The first one lasted roughly 300 years. It extended from the repulsion of Kubla Khan's second attack upon Japan in 1281 to the initiation of an unsuccessful war of conquest against Korea in the late sixteenth century. This period was not, however, entirely peaceful because it was by no means free from damaging interior conflicts.

The second and definitely more peaceful era began in 1603 when the Tokugawa Shogunat was founded, and lasted until 1853 when an American naval unit made an attack on the country. But since the Japanese were in no position to resist the American "black ships," and since America sought peaceful trade rather than war with Japan, the attack did not lead to a real war. It was not until 1894, when Japan attacked China, that the period of peace which began in the early seventeenth century came to an end. The length of this extensive peaceful era in Japan's history (it lasted 291 years) has probably never been matched by any other nation state. But it is clear that we cannot explain it by arguing that the Japanese are strangely immune to aggression. The warlike Samurai were not immune to it, and it would be folly to view the Japanese as possessing a "human nature" radically different from that of other nationalities. Modern history in fact suggests the opposite: after the Meji restoration at the end of the nineteenth century, Japan grew rapidly into a powerful industrialized society patterned after Western models, and the twentieth century has witnessed Japanese military ventures into Manchuria, Korea, and China—and the year 1941 is so deeply etched in the modern consciousness that it needs no underlining. In other words, Japan has in modern times earned the dubious distinction of matching the aggressiveness of France, Germany, and England in the course of the nineteenth and twentieth centuries. What must be pondered, however, is the fact that people who proved to be aggressive in the twentieth century had in earlier centuries proved themselves capable of achieving extensive periods of peace. Surely something more than "instinct" is involved.

Sweden has always had the reputation for being a country of warriors, especially during the Viking era and during the Thirty Years' War. Indeed, its history might be an excellent illustration of how aggressive tendencies are displayed in successive generations. But Sweden has not participated in wars since the early nineteenth century, and has enjoyed peace since then. The genetic heritage of the Swedes cannot be used to explain Sweden's current peacefulness.

WAR AND INTERNATIONAL ORGANIZATION

All the views about the origin of war and its inevitability which we have recounted here are extremely dangerous and in fact help to promote wars. By erroneously asserting that war is a necessity, such views make the promotion of peace much more difficult. Why should we work for peace, if "natural" necessity dictates the inevitability of wars? Why not instead be passive and do nothing to oppose the command of "fate"? Why not just crouch down when the bang comes? It is not until we have eliminated such escapist attitudes that we may intelligently analyze the actual cause of war.

Many of the wars that have caused humanity to suffer have been analyzed in detail by historians, who have tried to clarify the causes of them. Also the question of the origin of war in general has been investigated, especially by a number of Peace Research Institutes which have been founded in the course of the past decade. But even without a detailed account of the results of the P.R.I.'s, one can make the almost truistic pronouncement that war, as well as misery and starvation in our modern world, is due to the obsolete political organization of our planet.

Even if we can localize the problem by such a general diagnosis, we have neither solved it nor made even a serious attempt to do so. We need no intricate analysis of modern political life to understand that our international organization is inadequate. Certainly a great deal of hard work lies ahead if we are to determine how we can improve it. But the problem of international collaboration is obviously worth all the effort we spend on it. It is the most serious of all problems we face, for it centers on the oldest and most compelling of political questions: how shall we establish cooperation between the inhabitants of this planet so that we have a minimum of violence and misery and a maximum of abundance and happiness?

VI

The Politicians

In the preceding chapter we suggested that the precarious conditions in the world are due to the inadequacy of our present international organization. This proposal directs our attention immediately upon those who are responsible for that inadequacy, namely, the politicians. We must be especially concerned with those politicians who are weaving, with varying degrees of competence, the delicate fabrics of world policy. Indeed they form a diverse group. Some of them, their chests decked with gold braid and polished medals, are generals who have come to power by a military coup. Others are revolutionaries, riding on the crest of a popular movement they have created

themselves. Some are intellectual farmers who have organized victorious farm revolts, and last, but not least,
some are party leaders who are products of a party
machine within which they have come to power by
skillful negotiations. Of all these, some are reactionaries,
others radicals; some are benevolent patriarchs, others
brutal; a few are young, but most are old. However, by
definition they are similar in two respects: first, they have
been able to seize political power, and second, they have
been able to keep it.

Perhaps we should add a third characteristic common
to these men: most of them live by the rules of the political power game; they fear, more than anything else, the
loss of power, and the hope of maintaining power becomes the center of their lives. It is this concern, rather
than the interests of mankind, that determines for too
many politicians what moves they will make on the
checkerboard of international power politics.

Many of the leading politicians in the world have been
inspired by idealism at the beginning of their quest for
political influence and power. They have felt upset and
indignant about evil conditions and have decided to devote their lives to the construction of a better society.
Several of them have been familiar with prisons from
within and have seen their comrades and friends killed
there or in guerrilla warfare. They themselves have had
the luck or the ability to survive and have been fortunate
enough to witness at least a partial victory for the ideals
they have fought for. But there are at least an equal number of politicians who have traveled to power in a more

comfortable way. Like an enterprising businessman who starts with a small amount of capital and expands his business into a trust which dominates the world market, a politician is able to advance from the menial position of secretary in a local political club to a leader of importance in international politics. The procedure is similar for both the businessman and the politician: they both sell and buy and buy and sell. The one accumulates money; the other, political power. What matters is that the enterprise yield big profits. The irony is that, for many people, neither money nor power can provide complete satisfaction: when they have acquired a great deal of money or power, they do not see their life's goal as fulfilled, but instead become addicts of a passion to acquire still more.

It is sometimes hard to distinguish between these two types of politicians. A confirmed idealist may grasp power by compromising his ideals, and even the most opportunistic climber may have some ideals buried under a crust of restless ambition. It is always difficult to assess human motivation, and this difficulty is compounded when one tries to see behind the veil of political action. One reason for this difficulty is the modern "made in New York" political image. Using all their vast propaganda resources, skilled public-relations men shape the aspiring politician to fit the mold the public desires. The result is a clean, well polished family man whose honesty, sincerity, good judgment, and patriotism are beyond question. Of course Madison Avenue techniques form a double-edged sword: opponents of the polished paragon of virtue will hire another public relations firm to shape him into a heartless ogre, a seditious rascal, or an intellectual incompetent. Needless to say, the truth behind the images rarely reaches the public eye.

Yet very often the images do reflect essential truths, and no one can doubt that even the line that separates idealists from Machiavellists may become blurred by the very nature of the political power game. An opportunistic climber could not succeed in politics without sponsoring relatively consistent ideas, and a sincere belief in these ideas often develops from his public avowals. An idealist, in contrast, must seek raw political power, for without it his ideals can only gather dust in the closet.

Surely it is clear, then, that the greed for power is an occupational disease to which politicians easily succumb. And it would be difficult to count the victims of this, the coldest and most dehumanizing of all vices. Like all vices, the thirst for power has its rewards: to see people kneel down before oneself, to humiliate one's adversaries, to parade on a red carpet, to stand in the spotlight and receive the ovations of the people. Unlike those who have been addicts of alcohol or narcotics, the addicts of the power drive bask in the admiration of the people.

THE GUILD OF THE POLITICIANS

Politicians fight each other, sometimes by diplomatic procedures, sometimes with military weapons. Frequently, they make alliances with each other. The drama of world politics attracts so much attention that it is easy to forget that the politicians form a guild, the most privileged guild in the world. When they meet, it is not always to negotiate for the benefit of their people. Some of their consultations may be concerned with how to keep the power over their own people. Thus any solution to the common problems the politicians are trying to solve must be such that it strengthens, or at least does not weaken, their prestige and power in their own countries. Though they are representatives of their people, they also represent the politicians' guild, whose interests may be separate from the interests of their own subjects. In times of war they are anxious that the war should not be so "uncivilized" that it becomes disagreeable to them personally. They would prefer to sit and fiddle while Rome burns.

POWER: FROM GOD OR FROM THE PEOPLE?

In earlier times, rulers derived their power from some god. A great king, regarded as the direct offspring of a god, was believed to have inherited the power and the right to govern directly from his heavenly ancestors. His orders reflected the will of God, conveniently interpreted. The ruler delegated his power to his officers, and even the power of the least important of his servants issued ultimately from God. To obey the authorities appointed by the king was a religious duty. Today it would be absurd for a national leader to claim that he was ruling by divine right.

The Enlightenment philosopher J. J. Rousseau, who

contributed the most to the intellectual basis of the French Revolution, effected the fantastic achievement of replacing God, as the legitimate source of all power, by the people. As a consequence the family tree of a man in power today does not show that his grandfather's grandfather was a god. Instead it shows that his father, or at least his grandfather, was a simple man of the people. His power is derived from the people who have elected him as leader by giving him the majority in a general election. In a country ruled by a dictator, the majority is often claimed to be more than ninety-nine percent of the votes, whereas in democratic countries it need not be more than fifty percent, and sometimes even less.

In reality a modern ruler need not represent the people any more than rulers of earlier times represented God. The power of both types of ruler is founded on an organization which is controlled largely by the ruler himself. In earlier times it was more common for power to be seized by military means. Those who seized it organized a group with a king or an emperor at the top, surrounded by noblemen of different degrees. Such a group, organized into an administrative and military heirarchy, often wielded power for centuries. The group was self-recruiting; that is, it decided who should be admitted to its ranks.

There were also other power organizations. The church was another self-recruiting group, and the economic leaders formed a third. These groups collaborated in repressing the people but fought between themselves, especially on the issue of how to share the power. The masses had a minimum of influence, but had to pay the bill.

SELF-RECRUITING POWER

In the countries in which it was introduced, democracy brought, of course, much better conditions for most of the people. However, it is doubtful that it has changed the structure of power essentially, for the power still belongs to self-recruiting groups who govern or compete for political control. The earlier groups consisted of kings and noblemen, and those of today consist of party leaders and their party machines. Recruitment of the leaders today differs in some respects from that of earlier times. For example, power is not inherited to the same extent that it was previously. But the old Roman adage *mutatis mutandis* still applies: that is, the snake may have a new skin but it is still a snake.

In countries ruled by a single political party, the ruling group receives in an election an endorsement which is always manipulated to show a very large majority. In countries with many parties, the people have in theory the freedom to elect whoever they want to be their ruler. This is a noble principle, but unfortunately reality has not made room for it. Only those who belong to a party machine, and, of these, only those who have been approved as candidates have any chance to be elected. Recent events have demonstrated that a politician can enjoy widespread general confidence and carry the banner of very popular ideas that a great number of people want to see realized, but still have no chance in an election unless he is supported by an efficient party machine.

If one is to win an election, an enormous amount of money is often required and this can usually be raised only by the party machines. But why is money essential in a procedure which ostensibly functions to allow the people to elect a leader in whom they have confidence? The answer is that in an election campaign the people are subjected to intense propaganda that has been conveyed by the most sophisticated methods modern Madison Avenue has to offer. Only in part is an election a contest between differing ideologies. More and more it is becoming a struggle between different advertising agencies or similar firms. Such an organization accepts a job to sell a toothpaste or to sell a politician to the consumer, the people. Just as the face of a TV model advertising soap is carefully made up by make-up experts, so the faces of candidates, too, are cosmetically improved before they go before a television audience. The "image" is all-important, and today that image is being created with a great deal of psychological sophistication. It is sometimes claimed, and justifiably, that it is a publicity agency and not a party that is the real victor in an election.

Even if the power of the people in a democracy is rather weak, democracy nevertheless has many advantages. It provides more freedom to a large number of people. Power can be transferred from one group to another without bloodshed. The people feel that they themselves possess the power. Haven't they elected a ruler in a free election? Similarly, our freedom to purchase whichever toothpaste we wish gives us the impression of exercising free choice even if we are well aware that publicity and advertisements have actually decided it. It is easy to forget that in a political election the actual range of choices is very limited. In many elections, the political parties that are seeking power have similar programs. In others, the alternative that the majority really wants may disappear for one reason or another before the election. Then the election may offer a choice only between a disaster and a catastrophe.

It is of course comforting that leaders in a democracy enjoy direct contact with the people, but such contact is by no means confined to the democratic system of government. Harun-al-Rashid donned a disguise and sought to know the people's wishes as he walked the streets of Baghdad. More than a hundred years have passed since national monarchs, emulating Shakespeare's Henry V, fought side by side with their soldiers, risked all that they risked, and sometimes lost all that they lost. And, with all due respect to the well coached talents of today's smiling presidents and prime ministers, it is doubtful that any of them have surpassed the modern dictators in their apparent contact with the people.

Because the political leaders in reality are appointed by the party machine, it is of great importance to observe what qualifications the machine looks for in recruiting candidates. Certainly the candidate's intelligence and ability to work hard is very important. So are his abilities to discuss and negotiate, since these will enable him to obtain more power for the party. Servile obedience to those who have power already, and the art of maneuvering cleverly in the channels of political flattery and manipulation are also, unfortunately, helpful to the man who seeks to rise on the ladder of power.

THE STRUCTURE OF POWER

In spite of all propaganda to the contrary it is possible that the difference between ruler and subject is greater today than in earlier times. The reason is that modern states have a much more complicated organization. They are increasingly centralized, and furthermore the population is much larger now.

A tribal chieftain who ruled a few hundred men knew every one of them personally. They fought side by side, and when one of his followers was killed he often felt it as the loss of a personal friend. In the Greek states the total number of citizens rarely exceeded a few thousand and, because everyone knew everyone else, the leaders still had personal contact with a large portion of the population. But in today's nations, with millions or hundreds of millions of inhabitants, such personal contact between ruler and citizen is impossible. The people are necessarily grouped together in large numbers, which their leaders view as figures in statistical tables. Because of that, leadership has become dehumanized. For example, in his daily work routine, a business executive can have direct contact with five, or let us say ten, employees. If his organization employs a hundred people it must be divided into a couple of different levels. The director has, for example, ten subordinates, each being a leader of ten others. In an organization of a thousand people there must be still one more level between the director and the majority of the employees. The greater the number of people, the more numerous the levels between the chief director and the majority of the workers. Every additional level causes the chief director to become that much more isolated.

The problems which are vital to the life of the people are pushed from one level to the next. At each successive level, a man's individuality fades still more while the figure representing him becomes increasingly more important. When a particular problem finally does come to the attention of those in power, it remains unsolved, because at the executive level concern with the destiny of mankind has been obscured by other priorities, such as "economic necessity," "security of the country," "pride of the nation," and perhaps also by the power of the party and the personal prestige of the leader.

MEN OR STATISTICAL FIGURES?

The constantly increasing complexity of modern society contributes to this dehumanization of government. Every decision has a number of different and often unforeseeable consequences. Those in charge of the different levels in the structure of power have examined the matter, and given their opinions and recommendations. When finally the decision is made it is often difficult to know in advance what it really will accomplish. It is not until later, after it has been carried out, that everybody understands how it has affected the life of individual people.

A surprised despot, a few hundred years ago, is reported to have said, "Have We really done all that?" The democratic leaders of our time are very often equally surprised.

If a rural population is forced to relocate in a big city, difficulties in adaptation and much human suffering can be caused by the shift to an alien environment. In addition, a farming area may be turned into a desert. But those who are responsible for such fundamental changes see the people who have been displaced only as statistical figures, and are concerned solely with the possible economic benefits gained through the decision.

Measures to increase the number of births or to restrict birth control in a world that is already overpopulated have enormous consequences. Yet such actions are often taken, with little concern for all the tragedies they cause. Other decisions may forever shut the door of life for, to take an obvious example, the soldiers who are sent out in a war, or for those declared to be enemies, on whom bombs are falling like rain. A person cannot avoid asking himself how such decisions affect those who make them. The problems must look very abstract to them because they cannot envision all the human tragedies their decisions lead to. Apparently they are sleeping well. Perhaps they dream about statistical tables rather than the destinies of individuals.

It is evident that this political manipulation of the lives of men is no more cruel today than it was previously. But neither is it less cruel. This is shocking. In past times, when most people lived on the edge of starvation, the main problem was to survive. In some political situations the only alternatives were sacrificing a number of soldiers on the battlefield or watching perhaps still more people starve to death. Many times the question was not "Must we forfeit a life?" but rather "Must we forfeit this life or that life?"

When the Huns invaded Europe, they did so because their own country could not feed them. The Goths invaded the Roman Empire in order to save themselves from the Huns and to get more food. The law of the jungle prevailed: eat or be eaten.

But the cybernetic culture, which has provided so much material abundance, has fundamentally changed the circumstances. Modern medicine has made it possible to regulate the population. We have the technological means to satisfy all reasonable needs of a limited population. It is no longer necessary to let large numbers of people perish in misery or be killed in war. Yet reports of these very types of disasters continue to appear in every newspaper, but today such events are not due to the cruel necessity of nature but to the obsolete political structure of the world. The political games with the destinies of mankind are like those of ancient time, not because they have to be, but rather because too many of us think they have to be.

INCOMPETENT POWER GROUPS

In our fantasies, most of us envision the day on which the world's leading politicians will meet and draw up a global plan for ending war, effecting a general disarmament, and abolishing all hunger and misery. This day has not yet arrived, however, and very few really expect it to arrive in the near future. Why are these measures impossible? Aren't they precisely what almost all of mankind desires? The politicians claim that they want to do everything for the benefit of their people. Why don't they use their power to accomplish what the people most desire?

The answer is often given that contemporary world problems are so extremely complicated that only very naive people can believe that it is possible to solve them. To be sure, such spokesmen concede that we have the resources to produce enough goods to maintain a stationary world population, and the medical technology to keep the population stationary. Why, then, is it impossible to take the steps necessary to make the world liveable for everyone in it? The usual answer is that since the great politicians of the world, admired by many for the intelligence and skill they demonstrated while gaining access to power, are not able to solve the problems, this shows without any doubt how difficult the problems are. The conclusion is not necessarily correct.

In addition to the political leaders, another power group that is often expected to solve some of the world's most urgent problems is the military. For example, the French generals who served during the First World War were without a doubt France's most competent military leaders, because no one becomes a commanding general without having extraordinary qualifications. The generals believed that the war had confronted them with an insoluble problem. They thought that it was technically impossible to conquer the invading army, and this point of view was verified by the fact that the brave French army under the excellent leadership of the generals was not able to do it. The French statesman Georges Clemenceau, however, did not share this view. He believed that the actions of the generals were governed too much by their internal animosities and by obsolete military traditions. His opinion was that "war is much too serious a matter to be entrusted to the military."

An individual general must have viewed his position somewhat as follows: his most important problem must have been to maintain his prestige among his colleagues and his status as a brilliant strategist whom history would show to be the one responsible for victory. Secondly, he had to insure the interests of his men, and especially those of the military command, because if their interests were not protected, the whole military class might lose its power.

Of course he considered the welfare of his country to be very important, but the other two considerations still took precedence. One would naturally have desired the order of priorities to have been reversed. But, their gold stripes and medals notwithstanding, even the generals are human beings.

The world's leading businessmen constitute another power group. Their membership consists of a selection of the most clever people in the financial world. Intelligence, good sense, and energy are required to become one of them. For a long time they controlled the economy of the world. According to the economic doctrine of laissez faire, if everyone sought his own maximum economic well-being the result would be an optimum for the whole economy. Such a system however proved to be unstable. Booms and depressions alternated in a rather regular cycle. Many people asked whether this cycle was really necessary. Although the booms were encouraging, the depressions brought general misery to large parts of the population when the factories were shut down. Why could not the factories be put in action to produce all the goods the people so badly needed? The laissez faire theorists answered that the problems were very complex and anyone who suggested such simple solutions only showed how ignorant he was of the law of economic life. The fact that the most competent people in the world of business and finance could not prevent depressions showed that they were inevitable. It was regrettable that quite a few people starved to death during the depressions but very little could be done to prevent it.

In reality, a great deal could have been done. The problems were not all as impossible to solve as the leaders of finance and business claimed. It was the supposedly naive critic who was right. To apply Clemenceau's remark on the military to big business, one could say that the world's economy is much too serious a business to be entrusted to the men of finance and business. Government regulations have made it quite possible to eliminate the economic oscillations. The economic systems of the capitalist nations, regulated to varying degrees, still undergo certain oscillations but they no longer suffer from catastrophic depressions. In the communist world the fluctuations are under control.

Previously, economic catastrophes were caused by the economic necessities faced by every leading businessman.

First of all, he had to operate his own business and compete with business rivals. A failure to compete successfully could easily lead to his ruin. He had to sell when prices began to go down, meaning that everyone else was selling, and he had to buy when the prices began to climb, meaning that everyone else was buying. Second, he had to protect the common interest which he and his competitors shared. All of them had to unite to combat restrictive government regulations, demands for higher salaries from their employees, and other factors endangering the economic status quo. Only after these problems had been solved could the businessman devote his attention to the public interest.

No doubt it would have been desirable for the order of priorities to have been reversed. But, in spite of his wealth and social status, the businessman was after all only a human being.

As history can testify, the disastrous economic oscillations finally proved to be intolerable. It became necessary to declare the businessmen incompetent, and to put the economy under at least limited regulatory control by the government. Only by such control was it possible to avoid devastating economic catastrophes.

Let us now scrutinize the position of the politician. The individual politician is concerned primarily with retaining the power in his own hands and in the hands of his own party, with keeping his political adversaries down, and with upholding his image as the genial statesman, as the best possible leader of his party and his country. His second objective must be to defend the interests of his country against other countries, to protect these interests and his people, and to guide his nation to a dominating position in the world. Only after these goals are attained can the politician permit himself to work for the general welfare of mankind.

Surely it would be desirable if the priorities were reversed. But, in spite of the red carpets, the honorary guards, and the privileged treatment, the major political leaders of the world are only human beings.

WHO CONTROLS THE POLITICIANS?

If the generals or the businessmen are declared incompetent, this means that the politicians have put them under their control. This is possible only if the politicians have more power than do the generals or business tycoons. But if we come to recognize that world politics is too serious a business to be entrusted to the politicians, what then can be done?

According to the principles of democracy the people have the right to control the politicians. The people are supposed to do so in a general election by voting for those representatives they trust. But in reality this principle does not work very well. First of all the range of choice is limited to a few people whom the self-recruiting party machine has made its candidates. Second, as we have seen, the people have only a limited opportunity to evaluate candidates intelligently on the basis of facts rather than "images." They are subjected to the immense pressure of propaganda, especially at election time. Finally, often the blunt truth is that none of the candidates inspire or deserve much confidence or respect.

If it is difficult for the people to control the politicians, do other practicable alternatives exist? What about military takeovers, for instance? But who today would argue that generals who seize power through a coup d'etat are the ones we should entrust with finding a solution to today's complex problems? Industrial magnates can propose their own "solutions," but most of their proposals do more harm than good. The problem is that, whether it is a general, a powerful businessman, or a leading labor leader who grasps the reins of political power, he is still a man bound by the general rules of power politics. Boxed in by personal and national pressures which result from these rules, such a man will be unable to effect the changes needed in the structure of global politics.

Naturally we would wish that the major positions of world leadership were occupied by competent people whose vision was centered on the need to serve humanity itself rather than national interests. But it is sadly true that in many countries the process of political selection makes the election of such people extremely unlikely. Even if an occasional "philosopher on the throne" rises to power, he will have only limited opportunities to govern in a truly sensible manner. All politicians, it seems, find it necessary to plod along in the well worn tracks of the world's political systems.

We have witnessed some attempts to alter the structure of international relationships radically. Two of the most noticeable have been, of course, the establishment of the League of Nations and of its successor, the United Nations. Although these organizations have indeed aided in the solution of many crises, they have been powerless to effect the more fundamental changes needed to check the spiraling current of international tensions. The primary reason is that the United Nations is unable to challenge the moral, political, and military autonomy of modern national powers, which therefore remain free to dominate the world's political scene.

IS THE POLITICAL SYSTEM ACCEPTABLE?

We all hope that it will be possible to find, within the framework of the present political system, a solution to the world's problems; that the only action really needed is for the powerful politicians to focus their attention on solving the long-range problems of peace, disarmament, and abolishment of poverty. However, we cannot be at all sure that we are capable of solving the problem under the present system. Two centuries ago *l'ancien regime* of France was obviously incompetent and unable to transform the feudal society into a social order more suitable for the time. Likewise, neither Russia under the Czars nor Imperial China was able to adjust successfully to internal pressures for widespread reform. As a result, they, like *l'ancien regime*, were swept into the dust bin of history. All these establishments were overthrown by revolutions which introduced systems of a completely different structure.

The transition from an obsolete system, however, need not necessarily be accompanied by bloodshed. The feudal system, which the French Revolution liquidated in France, has in many countries, notably England and the Scandinavian countries, changed peacefully and gradually through a series of successive reforms. Hence if we conclude that the present world organization is obsolete and must therefore be replaced by a radically new one, it does not necessarily follow that a revolution must take place. Most analyses of the world's political systems center on proposed adjustments that might improve the present structure. But the most pressing need, surely, is to discover which organizational structures can most effectively promote human coexistence in the cybernetic epoch. It seems likely that a totally new structure is needed.

It is as remarkable as it is depressing that apparently no competent group of people is seriously at work on this problem, a problem on which the fate of the world depends. Most people seem to feel that human destiny is in the hands of the politicians, and that only the politicians can bear the responsibility.

But if the politicians prove to be incapable of solving our immense global problems, what then will happen?

The answer is that eventually another world war will explode.

VII

Man's Increasing Ignorance

KNOWLEDGE AND IGNORANCE

During the past centuries we have accumulated an enormous amount of knowledge. We know the structure of atoms and the distances to stars and galaxies. We have begun to understand the mechanisms of life itself. Psychology has given us some understanding of our mental life, and sociology of how society is structured. Newspapers report the events that happen throughout the world, and the Gallup polls tell us what everyone thinks of these events. At the same time our ignorance grows.

The reason our ignorance increases along with our collected knowledge is that we need so much knowledge in order to exist in an increasingly complicated world. Since our knowledge does not increase as rapidly as our need for it, the result is what we can call increasing ignorance.

Our collected knowledge is stored in libraries, many of which are organized in such a way that everyone has free access to whatever book he wants. Thus one would assume that anyone is able to make use of the enormous quantity of knowledge acquired by mankind during the course of centuries.

However there are three important requirements: (1) you must know what knowledge you need; (2) you must know in which book to find this knowledge; (3) you must be able to read. The second requirement is reasonably easy to surmount with the help of competent librarians. The first and third are much harder to overcome.

THE INCREASING ILLITERACY

The ability to read, of course, is not universal; a great number of people in the world are illiterate. In fact, the number of illiterates in the world is said to have increased by 200 million between 1961 and 1966. And though precise figures are not available, it is a safe guess that illiteracy increases by 30 to 50 million persons a year at the present time.

Although it is true that the number of people who can read is increasing (so that the percentage of the world's population that can read today is larger than that of, say, ten years ago), it is also true that the population is increasing even more rapidly. Consequently the total number of illiterates in the world is increasing. Why is this so, when it is possible to teach a young child of normal intelligence the basic reading skills within one year? It may, of course, take adults longer to acquire these skills, especially if they are holding jobs or raising a family, but they certainly can do so without great difficulty. Neither is it a great problem to train teachers and to organize schools.

SOME MODERN SUCCESS STORIES

The Soviet Union has provided a good example of how it is possible to eliminate illiteracy. One of the first steps taken by Lenin, after he had seized power, was to decree in 1919 that the entire population between the ages of eight and fifty should learn to read and write. The local soviets were given two months to plan and organize the massive effort. They were given the authority to mobilize all teachers not then serving in the Red Army, and the right to use churches, private houses, and factory assembly rooms for the purposes of education. This work was begun at a time when the regime was fighting for its very existence. Starvation and internal conflicts were ravishing the country. The civil war had not yet been decided. Soldiers were fighting with a gun in one hand and a spelling book in the other. A commissar would attach scraps of paper with letters on them onto the backs of soldiers marching in a line, and while they were trudging toward the front he would test the ability of those marching behind to read them. In many places they didn't even have paper or pencils. Teachers in these areas were advised to use the tips of charred sticks as writing tools.

The difficulties were especially acute in the arctic and

central Asian regions, where close to one hundred percent of the population was illiterate. In central Asia the difficulty was compounded by religious opposition which was especially aroused by the education of hitherto oppressed women.

After five years the campaign slackened somewhat, perhaps partly as a consequence of the illness and death of Lenin. But at the jubilee celebration of the Revolution's tenth anniversary, a new campaign was initiated. It was strongly insisted that it was the moral duty of every literate to teach at least one illiterate. Competitions between different villages and factories were promoted in the hope of encouraging everyone to learn to read and write. Literates were rewarded with citations and medals, whereas illiterates were put on the black list. In this way a national movement to eliminate illiteracy was so skillfully set in motion that nearly everyone sought to make his contribution. Peasants made blackboards and gave lessons in their homes, and whole villages joined the campaign to wipe out illiteracy. Once, when a peasant was asked how many illiterates he was teaching, he replied: "Not many, only four." "Why only four?" his questioner pursued. "Because everybody else is already studying," he replied.

By 1939, nearly ninety percent of the population had become literate. Today the Soviet people are perhaps the most "book conscious" people in the world. The country also has a higher percentage of skilled scientists and technicians than any other nation.

China claims to have liquidated illiteracy within a span of about ten years. Some Western sources have indicated that in actuality between forty and fifty percent of the population in China had become literate by 1968. Even if these figures should be closer to the truth—and it is not certain that they are—they still indicate a remarkable increase in the number of literate people in China since the 1950's.

The reason illiteracy in many new countries is so widespread and the number of illiterates is actually increasing must be that the rulers of these countries give this problem extremely low priority. It is even possible that some of those who are in power consider the spread of literacy, and consequently of knowledge, to be a threat against their regime. It is easier to repress ignorant people than educated people. The expectation explosion would be more violent if people who are now ignorant and suppressed could acquire the knowledge which would enable them to judge the competence and good will of their rulers.

WHAT CAN LITERATE PEOPLE READ?

Even to those who know how to read, most of the accumulated knowledge of mankind is closed. Many years of training at school is required if one is to learn to read major cultural documents, whether they are in philosophy, politics, economics or another discipline. In order to understand much of even one specialized science, a thorough university education is necessary. Therefore, even to the most highly educated people in the world, much of our collected knowledge is inaccessible. The way to the utilization of mankind's knowledge is through the specialists. We are completely dependent on them.

In an agrarian society a farmer actually had most of the

knowledge he needed, and consequently he was nearly self-sufficient. He built his farmhouse himself and could mend the roof if it leaked. He ground his corn in his own mill; his wife spun the wool she herself had cut from their sheep. He seldom needed help—perhaps only when he was ill or had to turn to the priest or the blacksmith. But today most of us are completely helpless without specialists and technicians of various kinds. Some of us can make minor domestic repairs, such as fixing a leaky water faucet or replacing a broken window, but few of us dare to take apart a refrigerator or a television set which is out of action. We do not know what these appliances look like inside or how they work. Yet this knowledge is not classified: in the library it is easy to find books which describe a television set's functional structure and components in detail. But how many of us are able to read these books? Our ignorance of the workings of items within our own homes has increased enormously, and this ignorance makes us helpless without the skilled hands of specialists.

We must trust the specialists because we are too ignorant to overcome our need for them. But if they cheat us, what can we do? It is not so dangerous if we have to pay a little too much to repair our car, or if we are talked into buying a new television set when all the old one really needed was a minor repair. But our ignorance in other fields is more serious.

It is nearly impossible to know whether our food and our medicines are good or bad for our health. We know that large quantities of poison are spread throughout the physical world by polluted water and air. Those who are in power assure us that the safeguards are satisfactory and that we are safe. We know that the stockpiles of atomic bombs are large enough to kill all of us. Mass production of new bombs continues in spite of this. We know that all atomic reactors produce extremely poisonous radioactive substances in constantly increasing quantities. The people in power tell us that the bombs are necessary to protect us from attacks, and that pollution of our natural resources is under the best possible control. But can we really trust them? Do they not speak for the same establishment which manufactures the poison, makes a profit from it, and wants us to believe that everything is going just as it should?

The more knowledge mankind gathers, the greater the individual's ignorance of essential facts seems to be. The most important question concerns international politics. Are international relations really handled as competently as possible? Are the global questions actually impossible to solve, or are obsolete political methods mainly responsible for the present situation? Our ignorance does not allow us to determine a final answer.

THE IGNORANCE OF THE RULERS

During one of the world wars, two allied great powers considered it essential to have a line of communication and transportation through a neutral country. They sent an ultimatum to the king of the country. He called his minister of defense and his generals. The king asked them if they had carried out their duties and given the country a valiant and unbeatable army, and they answered that they themselves as well as the army would gladly die for the king and repulse any foreign attempt to invade the land. The king was gratified and ordered the minister of foreign affairs to reject the ultimatum. The great powers moved into the country and, meeting no serious resistance, they rapidly approached the capital. The king didn't know this. Everyone in the country was terrified, but no one dared to tell the truth to the king. The reason was simple. In that country it had long been the custom to kick every messenger who brought the king unpleasant news down a high staircase of the palace. Because of that custom, no one was anxious to tell the king about the desperate situation. The ministers could not make any decisions for only the king had the power; but they understood that something had to be done, and finally they succeeded in persuading a foreign diplomat to request an audience before the king and tell him how the situation stood. His diplomatic immunity protected him from being kicked down the staircase. But by then the invaders were already in the suburbs of the capital. The anecdote is said to be true, and since it concerns an exotic country it is in accord with many people's image of oriental dictatorships. These people would maintain, of course, that such a situation could never arise in a Western democracy.

But are the conditions really so different in a democracy? Isn't much of the information that our rulers receive biased? Although this information comes, of course, in neatly packaged parcels, there is one fundamental difficulty: namely, that the information must serve two functions. First, this material must give the leaders the knowledge they need in order to act, and, second, it must give them the justification for those actions, so that they can make use of that information when they try to prove to everyone that they have acted in a wise and competent way on all occasions. In other words the information they want to receive must provide them with knowledge and prestige at the same time. These two functions are not compatible. During a war the military command sometimes describes defeat as an indecisive battle, or perhaps even a victory, in order to strengthen the men's morale and their own prestige. But if the generals themselves believe in their own victory report, instead of analyzing the situation as it really is, then they are truly lost.

It is possible that many rulers, even those in a democracy, are trapped in this way. They select experts who are to provide them with the necessary information, but many of these experts have the same general background, the same beliefs, and consequently same prejudices as the leaders themselves. An expert who is anxious to be promoted knows his chief's weak spots well. Reports which prove that the leader has committed a blunder may be discretely concealed, whereas material which flatters the leader by strengthening his beliefs and prejudices is presented by the "expert" whose hope for an expected remuneration surges. It is a general rule that power creates servility, and it applies not only in the palace of an oriental despot, but also in the corridors of power in a Western democracy.

We have already discussed the way in which an administrative proposal becomes increasingly dehumanized, as it acquires more and more statistical dressing in its progress from one administrative level to the next. At the same time, unpleasant facts disappear and political wishful thinking enters. Too many decisions seem to be made in this way.

Probably this is the only way one can explain many of the decisions made by present-day governments in critical situations. Although the considerations which have precedence within the inner circles of a government are what ultimately determine a certain decision, these motives are probably among the most carefully guarded state secrets. According to the democratic principle, a government is responsible to its people for its decisions. But the government with a well trained administration, including an efficient propaganda organization, stands a good chance of not being held responsible. It is only in more dramatic situations, for example during a war, that governmental mistakes become evident to everyone.

Let us consider, as an example, one of the Second World War's dramatic events. When Hitler increased his armaments the French were nevertheless supposed to have good reason to feel secure. They were protected from German attack by the Maginot Line, a modern wonder of the art of fortification, completely impossible to penetrate. When the war started, however, it became evident that the Maginot could protect only the southern part of the border. There was nothing to prevent the Germans from driving their tanks from the north, through Holland and Belgium, and then directly on to Paris. The French people had slept quietly behind a fortification which did not really exist. How was this possible? There must have been people who knew that the Maginot Line had an Achilles' heel. Or were there? Complete freedom of speech prevailed in France, as in all western democracies. Why had no one told the people the real situation? Could it really have been unknown to the generals and to the government? It is not our intention to analyze the conditions in detail. The fact is that the government of France acted as if it didn't really know the truth about the Maginot Line. This leads one to the startling suspicion that the French government and the French people were no less ignorant than the oriental despot about the actual threat to their country.

It is easy to find other equally disturbing examples. For example, did the United States government in 1941 know that its Pacific Fleet was vulnerable to Japanese attacks, a fact of great importance in view of the increasing political tensions in the late 1930's? Was there no one at the high level of decision-making who had read enough history to know that the Japanese had initiated the Russo-Japanese War by a surprise attack?

There is no reason to assume that leaders who make everyday decisions in peacetime are less ignorant than they are in wartime. But in so-called "normal times" their mistakes can be easily covered up.

GOVERNMENT AND JUDGMENT

Having discussed the question of whether the government leaders have access to unadulterated facts, we now proceed to the equally important question of whether these leaders are competent to combine and evaluate the facts.

During the period preceding the Second World War the most important task of the English government was to get a balanced view of Hitler and nazism itself. Did the nazis really pose a threat to the Western powers or was Hitler just as ridiculous as many people claimed he was? First Stanley Baldwin and later Neville Chamberlain held primary responsibility for making a judgment on the issue. About Baldwin, Churchill has written that he knew little of Europe and disliked what he knew.

Chamberlain was a businessman from Birmingham. It is said of him that he realized that Hitler did not behave like a businessman from Birmingham, and therefore he treated him as if he were from Manchester. Even if this is a caricature, it does indeed help to illuminate the intellectual basis for Chamberlain's actions both prior to and at the Munich conference, which Chamberlain probably believed had saved the peace. Only some time later did it become evident to him that there were people in the world who were even odder than people from Manchester.

With a frankness not at all common among politicians, Hitler had declared his views and discussed his plans in *Mein Kampf*. It is not quite clear from their biographies whether Baldwin or Chamberlain had read Hitler's manifesto. It is certain they did not read the German version, though they may have read it in translation. Had they really tried to look at the world as German eyes saw it after the Treaty of Versailles and the Weimar Republic?

The truth is that they probably did not have the intellectual tools such a task demanded. Neither of them were familiar with the German language, and neither had ever studied German philosophy or culture seriously. Nazism was repugnant to them. It was, to be blunt, simply impossible for them to comprehend and evaluate the Nazi outlook intelligently. Yet the most important task for both of them as British prime ministers was to evaluate accurately the threat Nazism posed to England and to other Western powers. This irony would perhaps be humorous were it not for our realization of what followed after Munich.

Their lack of judgment is completely excusable. No man can be reproached for not being able to accomplish what lies beyond his capacity. But the price England had to pay for its mistake of selecting incompetent men as its leaders was several years of hard war and the irretrievable loss of its world dominance.

But why dig up old bones? What is the use of discussing the mistakes made by politicians long ago? History is full of mistakes: we have all committed them, and we will all do so many times again. Why should we blame the world's miseries on some unfortunate politician who cannot be expected to possess superhuman qualifications and who, in reality, is doing the best he can to master extremely difficult situations?

But this is not the point. It is a dangerous oversimplification to say that everything would go well in the world if only all politicians were prudent and competent people. Rather, if we state that modern leaders often err, this claim should be taken, not primarily as a criticism of their personal competence, but as a comment on the present political system which puts too many ignorant people in posts in which they endanger the security of us all.

The development of worldwide communication systems has brought us into close contact with other people. We must learn how to coexist with them. But even today, many of those who have been entrusted with the task of solving the problem of coexistence are at least as ignorant about other governments as Baldwin and Chamberlain were about the Nazis.

How many wars have started, or will start, because someone misinterprets the needs or the intentions of people living in a culture different from his own? The gravity of such an occurence is evident.

OUR IGNORANCE ABOUT EACH OTHER

Especially during the period of the cold war one of the most important problems for the West was what type of relationship with the Soviet Union should be established. Was there a genuine risk of nuclear war? What were the actual intentions of the Russians? One of the most important tasks for Western diplomats was evidently to understand why the Russians acted as they did and to use this understanding to negotiate with the Communist bloc. Were the leading statesmen in the West competent to handle such a task?

The amount of knowledge the Russians have about Western countries greatly exceeds that possessed by Western countries on Russia. Many of the leading Soviet figures in the sciences, the arts, the army, and the government have a good grasp of English and often of German and French as well. Their knowledge often includes familiarity with Western literature, history and geography. But in the West, knowledge of Russian is quite rare among political and social leaders.

It would be interesting to find out how many leading Western politicians know Russian. The number is probably very small. How many know Russian history and have studied the evolution of Marxism so thoroughly that they can really understand those factors which underlie the thinking and actions of the Russian leaders? Certainly not many.

One may argue that a leader of a government does not need to know foreign languages because a sufficient number of translators and translations exist. Likewise, why should Western leaders study Marxism since there are so many experts who can tell them about Marx's teachings. But everyone agrees there is an enormous difference between an original and a translation. There is an even greater difference between listening to an expert review the conditions in a country and having lived in that country and inhaled its atmosphere. Furthermore the politicians look upon the experts with great skepticism. Information and recommendations must pass from the experts (who are often rather low on the totem pole) upward to the chief executive. In the process, those least capable of evaluating, say, the intentions of the Russians in an international crisis, will have more authority than the experts who are directly familiar with the Russian language and culture. Only rarely will the experts be allowed to participate when the ultimate decision in policy is made.

Naturally misunderstandings arise in the course of this process, and the unavoidable result is that Western statesmen consider the Russians to be unreliable and capricious when in fact these criticisms would better apply to their own chain of command. It is when such lack of communication prevails that Western ignorance may be far more dangerous than a so-called enemy nation.

Even more difficult are the relations between the West and China and Japan. A number of traditional prejudices and assumptions of superiority still strongly condition these relations. It goes without saying that the personnel at an embassy in London and Washington speak English, and in Paris, French. But it is not necessary that all ambassadors to Moscow speak Russian, and only in exceptional cases is the knowledge of Chinese or Japanese considered a necessary qualification for the people representing Western nations in Peking or Tokyo. Thus, a number of the people who carry the diplomatic responsibility for the increasingly important relations with China and Japan are illiterates in the country in which they work.

In the leading circles of China are many who know Western languages and are familiar with Western ways of thinking. Chou En-lai, for instance, has a superior command of English, and he also knows French, German, and, of course, Russian. But, none of his counterparts in the West, as far as we know, possess even an elementary knowledge of the Chinese language or culture. For these Western leaders, the Chinese ideograms—the graphic symbols of the longest continuing cultural tradition on our planet—are nothing more than curious scribblings.

To translate from English to French or German is quite simple. Translations from one of these languages to Russian are more difficult. To make an adequate translation from Chinese or Japanese to one of the European languages is extremely difficult, because these non-Indo-European languages have different structures, and their authors fill their writings or statements with allusions to their own classical literature, just as Western writers do. People who have not thoroughly absorbed the Chinese or Japanese cultural traditions are therefore often at a loss to understand the real meaning of a statement. To translate these languages word by word is not satisfactory, and very often it is completely inadequate.

It is as difficult for Chinese and Japanese people to learn Western languages as it is for us to learn theirs. It requires a great effort for them to speak a Western language fluently. This is not always understood in the West. One of the many deprecatory accusations brought against oriental people has been that they supposedly lack the ability to think logically. Such ridiculous statements can still be found in leading Western newspapers. But they stem from a fundamental ignorance of linguistics. It is only because the syntax of Japanese and Chinese sentences is quite different from ours that a literal translation of them can convey an impression of illogic. Yet, because many Japanese and especially Chinese public statements are indeed inadequately translated, millions of readers ridicule an "oriental mentality" which they have never understood. Needless to say, this misapprehension poisons the political atmosphere and obstructs a genuine appreciation of the Eastern cultures.

We all know that the iron curtain is said to prevent us from understanding the Soviet Union, and that it is the bamboo curtain which screens China from us. There is no doubt that both of these curtains are real. But we should not forget that there is another curtain which raises a still larger barrier to cultural exchange. This curtain is woven with the threads of Western arrogance, assumptions about cultural superiority, and thorough ignorance.

In past ages, when governmental policies were made by monarchs who inherited their positions, the future kings were given a careful education intended to qualify them for their high tasks. The most prominent scholars were employed to educate them in ethics, philosophy, history, administration, and other disciplines which they were supposed to know. The statesmen of today are not required to have such knowledge. By definition, they need to know only two things: first, how to seize the power; and second, how to keep the power. But the cybernetic age clearly requires a broader and deeper competence on the part of those who are governing our planet.

VIII

The Scientists

PARADISE LOST

The technological culture which has so radically altered the conditions of our existence originated from scientific research. This research and what it has led to continues to be the most important and extensive mental activity men have ever engaged in. Although the primary aim of science is to increase the storehouse of human knowledge, its secondary aim—the improvement of man's material conditions—has proved to be of immense and incalculable importance.

The avalance of scientific endeavor began with man's curiosity about the movements of planets in the sky and the influence of natural forces on the speed at which objects fall to earth. It was the study of such phenomena which gave birth to classical mechanics. Scientists today, building on the initial findings about gravity and planetary movement, are able to calculate the dynamics of machine operation, to measure precisely the rate of acceleration and braking in automobiles, and to devise routes for bombs, missiles, and space ships.

Another incredible leap of progress began when researchers early in the nineteenth century started to experiment with the tiny sparks which appeared when a person rubbed a stick of sealing wax or glass. Indeed this phenomenon ultimately produced effects more remarkable than those caused by Aladdin when he rubbed his magic lamp. These simple experiments mark the discovery of electricity, and the eventual transformation of night's darkness into the electric daylight of our illuminated homes and streets.

Still other early investigators studied the attraction between a pair of magnetized iron bars, and the understanding of this phenomenon has, within a short time, enabled us to harness that power which impels electric trains and washing machines. It is equally amazing that only seventy years have passed since an experimenter systematically examined the weak radiation from a radioactive substance and began the movement toward a potential nuclear catastrophe.

But it would be misleading to suggest that science originated and developed from only an impartial curiosity: that is, from a simple search for knowledge for its own sake. Even the cavemen knew that knowledge is power. And certainly the desire for power—for economic, political, and social power—has had a significant influence on the development of science. It was while the alchemists sought for the stone of wisdom, for a way to transform lead into gold and for the secret of immortality, that they laid the groundwork for what was to become modern chemistry. Also it was the desire to improve the quality of harvests and livestock, and thereby to increase economic power, which eventually led to the development of genetics, and the discovery of chromosomes and genes.

The importance of science to society began to be generally recognized in the eighteenth century. In that century scientific academies, dedicated to the improvement of various sciences as well as to the discovery of ways in which scientific discoveries could be practically utilized, were founded.

In the course of the nineteenth century, a number of important industries emerged as a direct result of scientific advances. Especially in Germany, the optical, electrotechnical and chemical industries, among others, owed their rapid progress to the discoveries made by scientists. Some scientists responsible for key discoveries also had crucial roles in the founding of industries for utilizing those discoveries. The immense industrial expansion of the nineteenth century, then, received great impetus from the direct participation by the scientists themselves in the industrial process.

The years before the outbreak of the First World War were in many ways happy ones for scientists. Sustained by the faith of the Enlightenment philosophers, who believed that any increase in knowledge must be to mankind's benefit, scientists were able to derive deep satisfaction from witnessing the practical utilization of many of their discoveries during this age of steam and electricity. At the same time, most of them were not so deeply involved in the industrial establishment that they needed to feel any guilt about its darker aspects: the exploitation of workers and the increasing squalor and misery in crowded industrial cities.

Certainly another contributing factor to these "good old days" for scientists was their relative economic independence. Most of their work did not involve great expense. Sometimes scientists were able to finance their own experiments out of their own pockets. Sometimes they were aided by individual benefactors. As a result, scientists were quite independent of the state. To be sure, occasionally a powerful man of influence would visit some of them to admire their home laboratories or to discuss philosophical issues. But, in sharp contrast to contemporary practice, such discussions did not have to end with the scientist pleading for a research grant.

FROM SMALL SCIENCE TO BIG SCIENCE

The "good old days" have been replaced by modern times, and the scientist's independence has undergone serious erosion. Many factors have contributed to this. One factor is the change in science itself. The days of the relatively simple experiments came to an end, largely because the phenomena which could be examined by unsophisticated and inexpensive instruments had been thoroughly observed. In order to delve more deeply into nature's secrets, more expensive instruments were needed. Like the rest of society, scientists had to relinquish the ideal of working as independent skilled craftsmen (just as the glass blower and the gifted blacksmith did) and join the complex of interdependent operations that constitute modern industry. One serious consequence of this shift was that scientists had to devote an increasing amount of energy to the procurement of grants. Although the "small science" of former times retained its own sphere of importance, the new "big science" could not function without large sums of money. Many scientists had to establish contacts with private industries or federal agencies, and use the flow of money from these for setting up laboratory units and purchasing needed instruments. This kind of work demanded considerable skill, patience, and tact.

The granting of funds to scientists is highly dependent on the good will and confidence of the granting agencies. In the United States, scientific research is financed through methods no less diverse than the society itself. Some support comes from private foundations, some from industry, and some from other sources too numerous to recount here. The bulk of financial assistance, however comes from the federal government. The advantage of the system in this country is that federal grants flow through a number of separate administrative arteries. Because of this, the possibility that an incompetent or unsuitable person in a key bureaucratic position will do much harm is significantly reduced. But, in the smaller nations, governed by a tightly knit administrative structure, the chance is much greater. In these nations bureaucratic clots may easily stop the flow of needed funds to scientific researchers.

In the Soviet Union we find a system different from that in the United States. The funding of research is controlled centrally by Akademiya Nauk which, like other academies of science in communist countries, performs the function that scientific academies were originally intended to perform. This sensible procedure should probably be attributed to the farsightedness of Lenin who, even in the early days of the Soviet experiment, gave the Akademiya Nauk considerable authority in helping to plan and direct the country's development. In spite of occasional conflicts with Soviet political leaders, the academy has managed to preserve its powerful position, and, although some mistakes have been made, it has on the whole managed the country's scientific and technical growth in a very competent way.

In Europe the planning and financing of research varies from country to country. A few countries have managed the job well, but most have unfortunately bungled it.

THE GEOGRAPHY OF SCIENTIFIC RESEARCH

A good example of the transition from individual endeavor to research within a complex industry is that which has taken place in the field of astronomy. If an astronomer has at his disposal instruments that cost, say a few thousand dollars, he can observe a limited number of stars and do a fair job of investigating the conditions in our nearest environment. But if his curiosity goes beyond our immediate environment, he must have larger and more expensive instruments. Millions of light years cost millions of dollars.

The United States was fortunate enough to foresee the importance of science to a complex industrial society. As a result, several large and well equipped observatories had already been constructed by the beginning of the twentieth century. Channels were opened through which money flowed to astronomers and other researchers, and organizations for the financing of research were progressively expanded. As a consequence, the United States took the initiative from her older European neighbors, and established a trend which gave tremendous impetus to scientific and the technological developments in many areas besides astronomy. We may perhaps perceive an analogy between this development and an earlier historical event that was of great importance to science. Galileo was born in Europe: and science thereafter became a European specialty. The great observatories were born in the United States: and the United States seized the reins of scientific-technological leadership. A similar pattern can be seen in the development of nuclear physics. As long as experimentation in this science remained relatively inexpensive, Europe was in a position to compete for leadership in the field. But when it became more costly in the 1930's, the United States again took the reins and assumed a dominant position in the development of this field, too. The vast economic power of the United States has therefore enabled her to establish what are almost monopolies in advanced aviation, computer technology, and the creation of new materials.

Aside from the United States, only the Soviet Union has realized the importance of taking the scientific-technical initiative. Sputnik was a dramatic attempt to wrest it from the United States. The instantaneous reaction of the Americans to the Russian achievement shows how well they understood what the stakes were.

For reasons too complex for analysis here, Europe seems to have definitely lost the lead. In spite of widespread destruction in Europe during the First World War, science was predominantly a European activity, even into the 1930's. The most important centers for the study of physics, for example were Berlin, Cambridge, Paris, and Copenhagen. But today most scientific research centers are located in Bonewash and Sanlosdiego (the urban "chains" of Boston-New York-Washington and San Francisco-Los Angeles-San Diego), Moscow, and perhaps Leningrad. It is difficult to tell how long Europe will attempt to retain even its present third-ranking position. Scientific research in Japan is rapidly expanding, and that of China should certainly not be underestimated.

KNOWLEDGE IS POWER

Science has undergone a natural change as a consequence of man's unceasing curiosity about his physical environment. But other factors have also contributed to its alteration. First the industrialists, and then the politicians, discovered that knowledge often equals power. Centuries ago, scientific discoveries were being applied to practical problems, and as we have seen, scientific work was sometimes inspired and supported by its potential technological application. It has often been said that the reason the science of the Renaissance led to a breakthrough while that of the Greeks withered on the vine was that the Renaissance discoveries found technological outlets. Greek science, mostly because of its close bond with philosophy, was kept locked in the study, so to speak, since it was only of limited practical use. Consequently it lacked adequate support from other groups in the society and thus shared the fate of classic philosophy when the new wave of Christian anti-intellectualism engulfed the ancient world. It is possible that the growing buds of science suffered a similar fate in India and China. But the sprouting science of the West during the Renaissance was more fortunate. It managed to acquire such influential benefactors among the leaders of economic and political life that the soil for the great breakthrough was sufficiently fertilized.

However attractive the infancy of Western science may have been, it is obvious that the extensive systematic exploitation of scientific achievements is a relatively new phenomenon that has effected sweeping changes in the nature of science. Scientific investigation has acquired status as one of the most important activities within society, and consequently, the forces which govern society have sought to incorporate science into the social structure. As a result, the position of the scientist has undergone a fundamental shift. Innocence and peace of mind have departed to make room for the burden of responsibility the scientist must assume for what he does, or for what scientists as a collective group do.

The systematic development of contacts between scientists and industry began, as we have mentioned, in Germany even before the turn of the century. This co-operation was based on the following principle: scientists agreed to place their discoveries at the disposal of industry, and to assist in converting them to marketable products, and industrial magnates in return agreed to pump back into scientific research a portion of the profits these products made. This system worked splendidly and many other countries began to emulate it. In the Netherlands and particularly in the United States, the German example has proved to be of decisive importance. But in many other countries the system has not functioned in an entirely satisfactory way for various reasons. In some countries, this failure may be traced to an inadequate understanding between scientist and industrialists about their mutual problems. Shortsighted industrialists prefer research which produces results that are subject to immediate exploitation, and thus fail to comprehend that many research projects must be long-range activities. Considerable time and money are required to organize and develop a research group. Unfortunately, the fact that scientific-technological initiative has immense value, quite apart from its immediately exploitable results, is regrettably incompatible with the ideology of the "efficiency experts."

In all modern industrialized nations, research has become an integral component of the industrial process. This means that the scientists must share the responsibility for the drawbacks of industrialization. Only a few scientists have realized the degree to which these draw-backs can cause scientific endeavor to be used to the detriment of mankind.

But the most significant change in the scientists' position came as a result of the Second World War. It has been said that in the distant future, when even Hitler's name has been forgotten, this war will be remembered as the one in which the atomic bomb was created. But, in spite of the incalculable importance the bomb has had and continues to have, it is possible that future generations will remember Project Manhattan (which led directly to the manufacture of the bomb) primarily as the first demonstration of the tremendous results that could be achieved by boosting a scientific project to the hilt. The postwar era has provided two more recent examples of such results: computers and space exploration. More can be expected.

We have already touched on the international political consequences of the atomic bomb, and we will discuss it again. But for now let us limit ourselves to evaluating the "secondary" effects of Project Manhattan.

One important result is that political leaders in both the United States and the Soviet Union—the two major "nuclear powers"—have seen irrefutable proof of what a valuable resource the scientists are in a crisis situation. There is no question that, if the scientists could be mobilized, they would be a power factor of decisive importance in a politically hazardous time. Thus it is no accident that, in both these nations, scientific research is considered of primary importance to the state. In addition, both countries have considered it worthwhile to build up a fund of knowledge in even those areas which are not directly useful ("useful" meaning those which yield immediate economic profit). Leaders in both countries have come to understand how impossible it is to foresee where a project of an essentially fundamental nature will lead. Also, since research is very inexpensive if considered from the perspective of the gross national income of a major industrial nation (for example, more money is spent on cigarettes than on basic research in the United States), the governments of these nations prefer to bet on promising research projects rather than to take the risk of strangling a development that could become important, perhaps very important.

But this attitude does not exist outside the United States and the Soviet Union. In most other countries it is maintained that the scientist's work should yield an immediate monetary profit, and this attitude has the effect of smothering the flames of long-range research discoveries. Since this trend in the smaller nations is unlikely to be reversed, we can expect the two "super powers" to remain the only research giants for a long while. Certainly this wholehearted support of research has diminished in the United States in the past few years. We can only hope that this is but a transient phenomenon.

THE DILEMMA OF THE SCIENTIST

But if Project Manhattan has significantly altered the attitudes of political leaders toward the importance of research, it has had an equally important impact upon the scientist's own attitude towards his work. It has caused many scientists to feel guilt and concern about the social consequences and political implications of their research.

Even before the twentieth century, scientists had been accustomed to producing results which were exploited by industrial firms and politicians, but they regarded their products as essentially conducive to the welfare of mankind. During the First World War, however, the new achievements of science included mustard gas, whereas the Second World War brought atomic bombs and missiles. It was with surprise and shock that scientists realized their activity had two sides. On the verso of the medal, they saw the sign of destruction. As a result, many of them have been caught in a serious conflict of conscience. It is common for the scientist who has worked on an interesting problem to find that the solution he has reached opens new and important vistas. But if he understands that a new discovery can be used for killing, destruction, and repression, what should he then do? If he wants to withhold what he has discovered this often means only that he postpones the dangerous discovery perhaps a few months, or at most a few years: only in rare cases is he the only one capable of making the discovery. What he has understood today, many of his colleagues will understand tomorrow, even if he withholds his results.

If, because of his scruples, he does allow himself to be scooped, his personal prestige may suffer and he may be unable to obtain the funds he needs for his laboratory, so that his opportunities to continue scientific research are severely curtailed. Also, his native country may be placed at a disadvantage if that discovery is made, and its military applications developed, by someone in another, perhaps hostile country.

The fact is, then, that a discovery which may be dangerous to mankind often gives substantial advantages to the discoverer, his laboratory, and his own country. These short-range advantages are usually given more emphasis than the long-range disadvantages. This imbalance is one of the most fundamental dangers of our time. One buys personal, local, and national influence at the price of a more dangerous world situation for everyone. The disruption of this fatal process is one of our most urgent, but also most difficult tasks. We cannot be confident that a solution is possible within the framework of modern political systems.

Many scientists meet such problems in their daily lives. If they are at all concerned with the moral aspects of their own work, it follows that they must be interested in the new milieu they create for all of us. But there seem to be only a few who demonstrate such a sense of responsibility.

The position of the academic scientist in his ivory tower is very comfortable. He often has a prestigious title and enjoys a protected life in a pleasant academic sanctuary. To involve oneself in the business of the world is to break an academic tradition. In many countries such involvement lacks academic prestige and does not contribute to one's professional advancement. Because of these conditions, most scientists tend to shift the blame for the negative applications of their work to politicians or businessmen, or to those who transfer the results directly from the laboratory to the practical world. They do not accept any moral responsibility for the consequences of their work; they have done their jobs.

WHO OWNS THE RESULTS OF RESEARCH?

But the problem also has another aspect which has not been widely recognized: large groups of highly competent scientists are forbidden to publish their results and to share conclusions that might be of great importance to all of us. For example, many scientists employed in industries are restricted in this way. If a scientist in an industry knows that a certain project which his company plans to undertake will poison nature, or harm or kill people, he cannot reveal this because his contract forbids him to do so. He has been hired by the company because he is useful to it, and he is not allowed to make any public statement without the consent of his boss. If he is the only one in the world who knows how harmful some of his company's projects are, then the boss may tell him to keep quiet. His knowledge is the company's property; he must watch silently while the economic exploitation of his discoveries causes harm to his fellow men.

Such dilemmas occur not only in private companies, but also in government scientific projects and government laboratories. In many of these, the scientists are formally forbidden to make any public statement without the consent of their superiors. Even if restrictions of this nature are not formally imposed, a public statement that

displeases his superior can have such disagreeable consequences for the scientist that he may prefer to com-
promise his conscience instead. Although government
organizations are supposed to serve the people, there are
many administrative layers between the people and the
director of any given organization. If there is a conflict
between the director's prestige and the welfare of the
people (for example, their protection from the dangers of
industrialization), it is possible that the prestige of the
director will be the decisive factor.

In Chapter Six, we investigated the personal situations
of the general, the businessman, and the politician. The
position of the scientist is sometimes similar. He thinks
first of his own career, and second of the welfare of the
institute or the company he serves; once again, the welfare of mankind must often yield to the two other priorities. It would be desirable for the sequence of priorities
to be reversed. But in spite of all his knowledge, the
scientist is only a human being.

Hence, the academic tradition plus the scientist's position in governmental and business organizations prevent
more than a minute fraction of our collected knowledge
from being available to the majority of mankind for protection from the dangers of the technological age. The
knowledge is at the disposal of those who have the power,
and there is little that the rest of us can do to shield ourselves from its misuse by those who control it. It is not
because of our ignorance that nature is polluted and dev-
astated. Many of these adverse effects have come as no
surprise to the scientists, many of whom were aware of
the potential dangers long before the deleterious actions

were carried out. The real cause is the relationship between power and knowledge in modern society.

During the Middle Ages philosophy was called *ancilla fidei*, "the servant of the faith." Today, science is the loyal and usually silent servant of power.

But there is no rule without exceptions. A small number of scientists have broken the silence and warned of the threat from atomic bombs, protested against pollution of the air and water and stressed the danger of the population explosion. By supporting their protests with the full authority of irrefutable scientific information, these men have created a sound basis for dissent from policies of aggression and exploitation. As a result of their dedicated efforts, a world-wide voice of conscience has been raised. To a large extent, it is because of them that no atomic warfare has taken place since Hiroshima and Nagasaki. Politicians and generals have often threatened to use the bombs, but protests from aroused people throughout the world have prevented them from carrying out their threats. About these scientists, we can certainly say that seldom have so many had so few to thank for so much. But their success has been only partial. They have not been able to prevent the atomic stockpiles from increasing every day, meaning that the danger of a catastrophe is also increasing. In spite of their outcries, the pollution of nature, too, is snowballing in a frightful way—as is the population explosion.

IX

The Population Explosion

THE INCREASE IN POPULATION

If every male and female couple of a species produce an average of two offspring, the population will remain constant in the long run. If they produce four, the population will increase eight times in three generations and more than one hundred times in seven generations. But if they produce an average of one, the population will decrease to one-eighth of its present size in three generations and to less than one percent in seven generations.

When members of the human race belonged to small tribes, it was the ambition of every tribe to multiply and to become numerous and consequently more powerful. Every clan chief rejoiced when his clan increased, for it meant an increase in his power. The nations who control the earth today are descendants of tribes that multiplied rapidly. Population groups who reproduced slowly either died out or remain now as only small remnants of their former numbers; the power of their rulers has melted away.

Hence the great nations of today have grown with a long tradition of population expansion, and consequently both the leaders and the people have an ingrown feeling about the importance of an increasing population. This attitude exists also in the smaller units, since a family or a clan rejoices when it is blessed with offspring and considers extinction a disaster. Similarly, most people want their own race to increase in number and in power. This common desire of individuals and groups to increase has culminated in the population explosion.

As long as large parts of the earth were uninhabited and could be cultivated, the population expansion was a boon for the human species. The domination by man became possible largely because the species became numerous. No culture can be developed by a group of people that is too small. Notably, the development of the cybernetic culture required an enormous amount of work which was available only because human beings were numerous. But the population has increased to the point of absurdity. Those regions which are not yet exploited are shrinking rapidly. The planet is full, or will become so very soon. Because the founding of a space colony is probably not a workable solution to the overpopulation problem, we must find elbow room on our small planet.

It is easy to calculate how much time will elapse before

the present rate of population expansion allows us no more than standing room for everyone on earth. However, such a state will never be reached, because other factors will have put a limit to the increase: one of them is the impossibility of producing food for that large a population; another is our limited capacity to dispose of waste. This vision of the future, then, reveals a hungry crowd of people on a plundered and poisoned planet, a condition in some respects analogous to that in a dying lake.

A number of objections to this vision of terror have been raised. It has been argued that large, still unexploited regions of the earth remain, and that we will soon have acquired the technological ability to boost our food production (the "green revolution") and to dispose of waste products. Some of these projections are no doubt correct, but it is irresponsible to seek comfort in them without emphasizing at the same time that they do not change the inevitability of the destiny depicted in the vision. In actuality, whether the earth is able to feed 10 billion or 100 billion or 1000 billion is irrelevant. It is true that our inventive ability and our organizational competence will conceivably enable us to accommodate the increasing numbers of people for a somewhat longer time. But the eventual result will be the terror we envision, because there is no escaping the results of exponential growth. To the principal question of whether or not the population can be allowed to increase, there is consequently one clear answer: that the present rate of increase is completely unreasonable. All the talk about finding additional resources in the world can serve no other purpose than to confuse the issue. No matter how much we increase our life-supporting assets, we cannot increase them infinitely.

The question of how many people the earth can accommodate is consequently of secondary importance, its importance being that it allows us to determine how rapidly the population must attain numerical stability, that is, zero population growth. If we can accommodate 100 billion people we have more time at our disposal than we do if we can accommodate only 10 billion. The length of the respite, then, is contingent upon the richness of the earth, but the principal issue—stabilization or no stabilization—remains unaffected.

MAN'S RIGHT TO REPRODUCE

Many people think that to have a family with many children is a cause for great rejoicing; and many never question their right to bring as many children into the world as they want to. But even larger numbers of people believe, and vehemently, that there exists a fundamental human right to have enough to eat. These two "rights," in a world in which food resources are not unlimited, must necessarily often collide.

At present an extremely large number of children are born to parents who do not want them. They are born because the parents do not know how to prevent pregnancy. It is therefore crucial, if we are to halt the population explosion, to overcome our ignorance of birth control methods, and to overcome the spiritual and secular powers which foster that ignorance. But the prejudices and taboos that these powers have managed to engrain deeply into man's thinking make enlightenment difficult. Moreover, enlightenment itself is not enough: it must be aided by economic measures and by vast and persistent public information programs on the importance of birth control.

Should the above measures prove to be inadequate, one might be forced to resort to more drastic steps: the possibility of "child rationing" looms on the horizon. Before anything so drastic is resorted to, it would certainly be wise to see what could be achieved by means of education—by the distribution of information and by massive efforts to persuade people that the population problem must be solved now. Yet in many countries such tasks have not even been begun, and what steps have been taken so far in other countries have been only halfhearted and therefore ineffective. No significant decrease in the rate of increase of the world's population has been yet achieved.

If we grant that every human being has the right to have enough food, we must insist that a social organization is acceptable only if it guarantees that every newborn child will have enough food throughout its lifetime. But it is an inescapable fact that parents who bring a new child into the world thereby place a burden on the whole of society—and ultimately on all mankind—and the parents must bear the responsibility that the introduction of this additional burden entails: that is, they must recognize that if the child is to eat his fill he must consume a part of society's, or the world's total resources, and these resources do not suffice for all of the world's present population. Before any couple decides to have children in an already overpopulated world, then, there may be good reason to ask: "Is there room for another person on our planet?"

People may disagree on whether the world's predicament is already so severe that we must ask ourselves that question now; or they may disagree on how much time remains before it will be necessary. But of one thing we can be certain: it is a question that will be urgent very soon.

POPULATION CONTROL

Without reservation, we can predict that the present population expansion will end within a few generations. It can end in one of two ways: either by mass death resulting from a global famine, a nuclear war, or a similar disaster, or by a rational limiting of the population growth until a stationary situation is reached. The first alternative will automatically decide the issue if we do not consciously apply ourselves to making the second one a reality.

It should be remembered that wars of the type we are "accustomed to" are of negligible significance in population reduction. All the victims of the two world wars were not more numerous than the present increase in world population in the course of half a year. Hence in order to limit the population by war we need two world wars every year. Famines, too, so far have never claimed more victims than one month's increase in the world population. If the population is indeed limited by a catastrophe, then, such a war or famine will be of a magnitude never before witnessed by mankind.

If we accept this first alternative as inevitable, we imply that humanity should imitate the life cycle of mold— population explosions followed by mass death. Let us therefore discuss instead the more sensible alternative: limitation of the population increase by birth control. How good are our chances of effecting a limitation by this means? Given the realities of the modern world situation, we must be careful to distinguish between what is technologically possible and what is politically possible.

TECHNICAL AND POLITICAL POSSIBILITIES

When we speak of a technical possibility we mean that a certain measure is practically feasible, provided that the world is organized in a rational way, with the welfare of everyone as at least one of collective humanity's main goals. Such a measure would not, in other words, require resources and knowledge which we do not possess.

The term "politically possible" refers to measures which can be taken within the framework of the present political system.

It is often claimed that only politically possible measures are worth discussing; that only a utopian can afford to waste time talking about what would be possible within a political system which does not exist at present. This is a sensible claim if we assume that the present political system is eternal. But many apparently stable systems have collapsed, such as *l'ancien regime,* the system in France before the Revolution. Its fall was predicted and in fact caused by the Enlightenment philosophers, who analyzed the theoretical deficiencies of its outdated structure and compared the existing opportunities with those that would be possible in a utopian system. It is therefore important to us, also, to discuss solutions which are technically possible, even if in the current situation they are politically impossible.

Let us first discuss the question of whether it is technically possible to halt the population increase. The population continues to increase because there are more births than deaths every day. We can prevent the rise in population either by increasing the deaths or by diminishing the births. Only measures of the second type are realistic.

Childbirth takes place because an egg cell has been fertilized by a sperm cell. If, starting today, we prevent spermatozoa from fertilizing egg cells, no more births will take place after the ninth month from now. Because this would not essentially affect the number of deaths, the population surge would stop and the number of people would begin to diminish. The requirement for the conjunction of a sperm cell and an egg cell is sexual intercourse. Since it is neither possible nor desirable to deprive people of the positive values of sexuality, sexual intercourse cannot be prevented. However, medical science has discovered a number of methods to prevent intercourse from resulting in conception. Several of the methods entail very little interference with the normal life of the people in question. Consequently, if they could be universally applied today, the population increase would stop in nine months.

If we were to make use of these methods to halt the population increase, the first step would be the production and distribution of sufficient quantities of contraceptives. Without having a detailed understanding of the technical problems involved, we might guess that a sufficient number of contraceptives could be produced and distributed within about one year. Add to this year the nine additional months, and we find that a period of slightly less than two years is needed to stop the population explosion, given our present medical knowledge and industrial capacity.

A second step would be the education of the population on the use of contraceptives. This is a simple problem, and rational organization and planning could solve it in a short time. If we compare the difficulties in using contraceptives with those in the use of modern weapons, we find that the second type is much more difficult. And yet we know that in a critical situation a large part of the population can be trained to become skilled soldiers in one year or less.

The main obstacle to the use of contraceptives is said to be the psychological resistance of the population. But there is also a great psychological resistance to learning the use of weapons. In spite of that resistance, the leaders of a country usually have little difficulty in mobilizing the population for warfare in a critical situation. The power and the propaganda machinery which modern governments normally possess are sufficient in many situations to make the people accept tremendous sacrifices.

The people can be taught to kill, to spend long stretches of time in trenches, and to subject themselves to all the horrors of a war. Surely difficulties involved in making people use contraceptives are trifling compared to those faced and overcome by governments in wartime.

THE NUCLEUS OF THE POPULATION PROBLEM

We can now approach the nucleus of the problem. If a government decides that its power is threatened by internal or external enemies it mobilizes its resources with an often unlimited brutality. But if the whole future of mankind is threatened, as it certainly is because of the population explosion, most governments hesitate to take the urgently needed measures even if these are very simple. Even if the population in a country grows so rapidly that famine occurs, this is a catastrophe that the power holders accept calmly, provided that starvation does not touch them personally. And it seldom does.

The population explosion is not a consequence of an irrevocable law of nature or of any other inevitable process. It is technically possible, even technically easy, to bring it to a halt in a couple of years. But the people in power do not view it as an important problem. Many of them—and this is what is really serious—hope for an increase in their country's population, though not in the world's. They are of the opinion that such an increase will enhance their power, and they are especially afraid that their power will diminish if the people of other nations increase more rapidly than their own people. It must be this fear which underlies the propaganda condoning population expansion, which both secular and spiritual leaders impose upon an already overpopulated world.

THE JAPANESE EXAMPLE

Our statement that it is possible to stop the population explosion within a few years is not unrealistic. We need only look to Japan for proof. After the end of the Second World War the leading circles in Japan considered it both necessary and advantageous to limit the population of the country. In fewer than ten years Japan's birthrate decreased from one of the largest in the world to one of the smallest. This was accomplished by abortions and by extensive and persuasive propaganda in favor of birth control. Many of those currently opposed to regulating the population describe the methods for doing so as "brutal," and have tried their best to find cases in which neurosis and other disagreeable problems have resulted. But such consequences must be compared with what the Japanese had to endure during a very brutal and devastating war. Many analysts agree that Japan's imperialist expansion was associated with the rapid increase in population, which forced the nation's prewar leaders to seek space to accommodate the expanding numbers. One should also note that in the decades since Japan stabilized its population she has enjoyed the most rapid rise in prosperity which any nation has ever been able to report. The causal relations between the stabilization of the population and the increase in prosperity are perhaps difficult to analyze, but the Japanese example does contradict the thesis that birth control is fatal to the future of the people. However, it is necessary to mention the sobering fact that in spite of Japan's present low birthrate, her population is still increasing in a way that many find alarming.

THOSE WHO PROFIT FROM THE POPULATION EXPLOSION

It seems likely, then, that the population explosion is caused by those secular and spiritual rulers who do not want to stop it and may even ardently wish it to continue. Their underlying motivation is probably the same as that of the old tribal chiefs. They consider it prestigious to rule a large number of people, regardless of what their standard of living is. If given the choice between a small group enjoying a high standard of living and a large crowd of poor and uneducated people, many leaders would prefer to govern the second group. One possible reason is that it is technically easier to rule poor and ignorant people than educated, prosperous people. Finally, many businessmen consider it essential to have a good supply of workers. As a result, they find mass production of poor people useful, for it gives them an easily exploitable reserve of cheap labor.

Spiritual leaders have more power if their followers are numerous and docile. The virtues they emphasize are humility, obedience, and acceptance of poverty. The larger the number of poor and obedient people in the congregation, the less the religious authority of the spiritual institution is questioned.

The most common, and dangerous, way for a leader to meet the demand for limiting the population is to admit that it is justified, and then to permit the introduction of a few token precautions which soon allow the population surge to continue as before. It is then possible to counter all criticism by claiming that the problem is too difficult to solve, an assertion that is simply not true.

Secular leaders are often anxious to increase the population because they are afraid of a hostile country or because they want to be in a position to threaten their neighbors. A large population and a rapid increase in the population have often been utilized as the rationale for expansion or imperialism. One arms by producing both guns and manpower to handle the guns. Too often such inhuman and cynical views are at the root of propaganda for population expansion or for condoning apathy about its control.

X

The Symbiosis Between Man and Technology

IS OUR SITUATION IMPROVING OR DETERIORATING?

In the preceding chapters we have tried to analyze the remarkable form of life which has evolved on the third planet.

A species appeared and developed itself to the degree that it began to analyze its own situation. It has begun to consider itself as a product of atomic and cosmic forces, and has used its acquired knowledge to control nature. Man is master of the atoms and is on his way to becoming the equal of the cosmic forces. He has had enough intelligence, then, to actualize all of these accomplishments. But the problem of organizing the coexistence of man and man seems to be too much for him.

What is the future of man? Let us imagine a dialogue that is taking place between a pessimist and an optimist on just that question. The pessimist will stress above all the horror-filled future prospects of population explosion, and the increasing quantities of deadly atomic bombs, other weapons, and poisons whose mere existence increasingly threatens the extinction of mankind itself. He will point to the polarization of rich nations and poor nations and to the gap that constantly widens between them as the rich grow much richer while the poor grow only a little richer, if not poorer. Finally he will point to our spiraling ignorance, and to the suspicion and hate it creates, designating them as both symptoms and causes of our dilemma.

But an optimist will answer with an equally long list of radiant prospects. The threat of the population explosion can be neutralized because we have the technical capacity to stop it. Similarly, he will maintain that it is technically easy to stop the production of atomic bombs and to destroy those that already exist. Our ignorance of our situation and of each other is countered by the fact that the total accumulation of our knowledge and competence is growing. With so many positive prospects, the optimist can regard our total situation as rather hopeful.

But if we scrutinize both lists we discover that the pessimist concerns himself mostly with what is happening in the "here and now," whereas the optimist is thinking more about what we could do, and not enough about what we are doing. It is true that man is starting to practice family planning; but some of the methods now advocated are inconvenient, and in many countries the national efforts to promote the use of birth control are limited and continue to meet strong opposition from reactionary forces. It is true that negotiations for disarmament are going on, but they are only a facade behind which the armaments race continues or even increases, partly because of the pressure from military-industrial complexes in many different countries and perhaps in larger part because of fear, hatred, and ignorance. It is true that the rich countries give aid to underdeveloped countries, but it is often of a type that does not provide any real help toward solving long-range problems.

If the same trends continue—namely, spasmodic steps of progress accompanied by serious deteriorations—the inevitable result will be a continuing degeneration of the world situation. This appears to be the state of the world today. We are probably sliding downhill. Why? The only possible answer is that under the existing political system it is impossible to solve the great problems of the cybernetic era.

GLOBAL STABILITY

It is possible that within the present governmental structures we could find the means to direct evolution more productively, and certainly all such possibilities should be utilized. But it is also possible that the present situation is fundamentally unstable.

Physicists know that, for many complicated systems, criteria that must be satisfied in order to insure the stability of those systems can be found. There are numerous examples of this type of criterion. If a balloon is filled with only a small amount of gas, it is stable in the sense that if we make a small hole with a pin the gas in the balloon will leak out slowly. There is time enough to mend the hole and to prevent the gas leak. But if the balloon has been blown up to more than a certain limit, which can be calculated theoretically, it will suddenly explode at the touch of a pin. Although analogies are always dangerous and never prove very much, we might say the balloon represents the political structure in the world, and the gas represents raw power and the means of destruction. If, in a structure capable of functioning with moderate power resources, more power than the structure can safely absorb is pumped in, the possibility of an explosion will be strong.

To offer another example, in plasma physics (the physics of ionized gases) scientists have studied the ways in which large numbers of molecules, atoms, ions, electrons, and photons interact. Such a "manybody" problem is extremely complicated. Considerable work is devoted to determining the conditions under which a plasma is stable. Because so many reactions are possible, the plasma can become unstable in a number of different ways. According to a recent publication, there are thirty-two different types of instabilities in a plasma. The denser the plasma—that is, the more particles there are within a certain volume—the more easily instabilities will arise. But the purpose of plasma physics is not only to map the various instabilities; this is only the first step. Its most important practical task is to cure them, to demonstrate how to make a stable plasma. By a systematic elimination of one instability after the other, the physicist tries to make the plasma stable.

The collaboration between a large number of people is

a problem which is immensely more complicated than the interaction between molecules, atoms, ions, electrons, and photons. But nevertheless it may be possible to find some general laws under which this collaboration may become possible. Many mathematical physical laws are so general that they can be applied to radically different problems. In principle it is possible to find a certain analogy between a manybody problem, such as that encountered in plasma physics, and the manybody problems faced by mankind today. This analogy, too, is risky, but using it as a basis, we might make the following conjectures about man and his world. (1) The larger the population on the earth, the more difficult it is to establish a stable condition. (2) If a certain population figure is exceeded, it will no longer be possible for stable conditions to exist. Rapid communications increase human interaction and hence human instability. A technical and cultural outlet in the form of a space colony, or perhaps several space colonies, would have a stabilizing effect. (3) Increased centralization has a stabilizing short-term effect but a disrupting long-term effect. (4) The most difficult question is whether it is possible to find stable solutions within the framework of the present political structure.

STABILITY AND HUMANITY

As we have already stated, the foregoing speculations are only guesswork. A closer analysis of complicated problems often yields very surprising results. Moreover, in an analysis of a manybody problem, we cannot expect to discover more than a limited truth, although it may be an important one. It is particularly important to remember this fact when we are dealing with the problem of human collaboration, because this is not exclusively a manybody problem; it is even more a manysoul problem. There is a great difference between atoms and men.

Furthermore, the problem of human coexistence is not only a problem of stability. Suppose that we have found a solution that does indeed enable us to avoid war and famines. This does not mean that the problem of human coexistence is solved. An equally compelling demand of the world's organizational framework is that it must offer men an opportunity to lead a worthy existence. The present trend toward dehumanization must be stopped.

No matter how we approach the problem of man's global coexistence, it is impossible not to recognize that the obstacles to its solution are enormous. But what is more serious is that no one group has seriously undertaken an unbiased scrutiny of the problem. Some aspects of the problem have been dealt with by U.N. organizations, others by various institutes for peace and conflict research. For example, an interesting and important activity is that which has been conducted by scientists from all over the world who have gathered at the so-called Pugwash conferences to mobilize opposition to the arms race. But it seems as if the problem has hardly been treated in a sufficiently penetrating and broadminded fashion. Most of today's investigations confine themselves to the particular question of what can be done to improve the present political organization. This question is important, and all efforts toward improvement of the existing political structure must be welcomed. But there are many other basic problems that are even more important. One such issue is whether the present structure, based on sovereign nations grouped in rival military and economic blocks, is competent to cope with the cybernetic age. Another is whether new methods for selecting and electing our leaders would be preferable to those we have today.

These questions have been widely discussed, but seldom have their implications been fully explored. Most people prefer to crawl behind a smoke screen. Or they comfort themselves with the hope that the great problems will solve themselves as the time passes.

THE USE OF REASON

Of course the major problems will solve themselves, but how? Are there any somewhat similar situations which we should study? The immediate answer must be no. For the first time in its evolutionary history, the entire biosphere of our planet has become one unit, with man exploiting all of it and directing its evolution. But the biosphere as an ecological system may be similar to a lake. We know that life in a lake develops as the result of an interplay of a number of autonomous processes, and that this development follows certain general laws. The vegetation, sparse when the lake is young, becomes more and more abundant. Life promotes itself until a certain stage is reached: the population becomes too dense, and the waste products accumulate. The lake grows smaller and more shallow, until finally the same autonomous processes that fostered the development of life in the lake not only cause the extinction of that life but of the lake itself.

We can see that the first phases of the evolution of the biosphere were similar to those of the evolution of a lake. When life started on earth, it made the earth more inhabitable: the oxygen in the atmosphere is a product of early anaerobic life and it is necessary for the more highly developed life forms. After populating the oceans, life invaded the land, which was originally sterile. Plants grew on the continents, making them suitable for animal life, and when man came along he exploited both plants and the other animals. Life promoted life, and the biosphere expanded to cover most of the planet. From the beginning of his activity man contributed to this expansion by cultivating the land.

But soon the optimum was passed. Even the early food-raising people so exploited the soil they cultivated that it became less fertile and much of it was eroded away. This may have been a main reason for the decay and fall of several ancient cultures. Today there are a number of processes which are dangerous: besides soil erosion, we have pollution, the accumulation of destructive bombs, and the population explosion. To a great degree, all these processes are autonomous, uncoordinated and, so far, essentially unchecked. To a terrifying degree they are analogous to those autonomous processes which cause a lake to enter the last phase of its life. Neither the lake nor the biosphere has any instinct of self preservation. An ecosystem is not a species possessing the intelligence to coordinate the different processes, and prevent them from interacting in a way that can lead only to a catastrophe.

However, one important difference exists between the two. The biosphere is dominated by man, who possesses enough intellectual capacity to guide himself, as a species, by the use of reason. He has already begun to recognize the danger, and it is quite conceivable that he can change his destiny by diverting or stopping these autonomous processes that threaten his future. Most people want these problems to be solved, and numerous declarations, promising that the problems will be solved, have been made. Moreover, there are many scientists, social scientists, and other professional people, with enough general training in the study of difficult problems, who could perhaps establish a viable plan for world organization and collaboration. A carefully planned framework which could satisfy man's reasonable demands for freedom from war, misery, and repression is what these people could conceivably achieve. Yet no serious attempt to devise such a framework has been made: apparently, man prefers to live as mold does.

It has been argued that even if it were possible to find a theoretical solution, there would be no chance to apply it in actual life. Suppose, for example, that the solution called for reducing the power and so-called sovereignty of the present politicians. The politicians would never accept this suggestion, and since they hold the power, no such action could be taken against their will. The dreams of starry-eyed utopians have very little relevance in practical life.

Although this objection is certainly very serious we cannot allow ourselves to despair. If it is inevitable that the present political situation will lead to catastrophe—and it seems to be inevitable—then any alternative possibility must be explored.

There are many historical parallels to our present situation. Italy, during the fourteenth and fifteenth centuries—the time of the condottieri—was perpetually unstable. Disastrous wars occurred in rapid succession—a succession more rapid than occurs today. In the history of China, there is a periodicity of a few hundred years. Strong centralized power enforces peace for one or possibly two centuries, and these are followed by a slow disintegration toward general anarchy. This ends in a new era of peace, enforced by a central power which again slowly disintegrates.

The economic instability that occurred in the totally unregulated capitalistic system is another interesting historical parallel, which we discussed in a preceding chapter. Booms and depressions followed each other in rapid succession until government restrictions changed the whole system.

It is possible to show a similarity between the regular sequence of economic catastrophes and the succession of political and military catastrophes which we witnessed in the world wars. We have had two such wars, and many people are afraid that we are on our way toward a third one. The political-military situation in the world seems to be unstable: an armaments race results in a world war, which is followed by a period of war fatigue, which in turn is followed by another armaments race. Is there any possibility of breaking the sequence?. The series of economic catastrophes wrought by depressions could not be averted until a theoretical analysis of the possibility of stabilizing the economy had been made. Government measures could then be taken to carry out successfully the proposals of the theoretical analysis. But the problem of global warfare is much more difficult. First of all, man has not yet discovered any workable theory for avoiding world war. Second, the solution of economic depressions presupposed the existence of a power which could enforce the needed measures. It was possible to disrupt the sequence of economic disasters because the politicians in the name of the people could put the leaders of the economy under their control. But if we are to stop the sequence of military and political disasters, where do we find a power that can control incompetent politicians? This is the central question, our fateful question.

OUR FATAL PROBLEM

It would, however, be unfair to blame the present world situation on the incompetence of the politicians. We all wish our leaders were more competent, had broader vision, and, above all, gave the welfare of the whole of mankind the highest priority. We might also wish that all persons in responsible positions were wiser and more sensible than they are. But the leading politicians themselves are caught in the revolving door of politics, and the mechanism of political selection functions in such a way that it gives power only to those who have the qualities required to seize it and keep it. Thus only through a stroke of luck does it happen that a person who comes into power has such qualities that he can and wants to use his authority to serve humanity.

We cannot expect our present situation to improve

without first analyzing it in detail and clearly determining which ways can lead to improvement. This is not a task that politicians can be expected to perform. Only rarely have politicians themselves been the sources of constructive new thinking of a fundamental sort. Rather, when politicians usher in sweeping changes they do so by putting into practice the systems that philosophers and scientists (both physical and social) have formulated. The inspiration for the French Revolution was provided by the Enlightenment philosophers; that for Lenin's and Mao's revolutions by Marx; and that for Roosevelt's New Deal by his brain trust, which consisted mostly of professors of economics.

What makes our present situation so precarious is that no one seems to be applying much extensive systematic thought to the problem of human cohabitation in the cybernetic age. We might perhaps express our present condition in this way: the utopian tradition is dead, or at least it is withering on the vine. The Enlightenment philosophers dreamed about a new and better world, as did Marx and the other socialist thinkers; but modern social prophets offer horror visions like that of George Orwell's *1984*. They talk about what we fear, but not about what we wish and hope for.

The predominance of this type of prophecy is probably a result of our shift from a symbiosis with nature to a symbiosis with our own technology, which has greatly weakened the value of old traditions and created a need for new ones. The traditional political and social ideologies cannot be expected to provide constructive solutions to the problems of the cybernetic age without a powerful injection of scientific and technological thinking. But not until very recently have scientists and technicians begun to take an interest in the social and political questions. Before any work of great value can be accomplished, there must first be established international groups from all disciplines who are free to cooperate across national borders without pressure from politicians. Should these groups formulate any concrete proposals, the most difficult goal would still remain: namely, to put them into practice.

It is a long and laborious path. But it is the only possible route, and it is essential that we take the first step soon.

XI

Are We Unique?

LIFE IN OUR SOLAR SYSTEM

In this concluding chapter let us return to a subject we touched on in Chapter One—the possibility of life elsewhere in the cosmos. Our brief visit to the third planet is ending, our time scale will be expanded again. The units of time will no longer be the seconds or microseconds that signify the hectic pulse of human or cybernetic life, but rather the thousands, or even millions, of years that denote the slow rhythm of the cosmos. We are leaving the earth not to forget what we have learned there, but because we believe that we need to look at it from a cosmic perspective in order to understand more about it.

If we pose the question, "Are we unique, are we alone in the universe?" we are asking what is technically an astrophysical question: we are asking if, somewhere in space, other civilizations like our own exist. In reality, however, we are asking a very important question about ourselves, or rather a number of basic questions about life, about its biological development, about the history of mankind, and about our society and technology. Hence the question "Are we unique?" gives us an opportunity to summarize the content of the preceding chapters.

A large number of science fiction novels are stories of earth men who come into contact with beings from another part of space. Many people have read so much science fiction that they believe they themselves experience what the books have described. They claim to have seen flying saucers and unidentified flying objects in the air, and some have even reported the appearance of small green men who smell like sulfur.

The question of whether life exists on other planets is very old but still undecided. Space research has determined that the probability of finding life elsewhere in our planetary system is not very strong. The moon is permanently subjected to a recurring process of sterilization by solar radiation. Once a month its surface is heated by radiation to more than several hundred degrees and it is simultaneously exposed to intense radiation or ultra violet light, X-ray radiation from solar flares, and cosmic radiation. We know now that no life of any type exists there. Venus has a very hot and dense atmosphere, which makes life impossible. Mercury is too hot. The existence of some lower forms of life on Mars is perhaps not altogether impossible, but it is unlikely. The outer planets and their satellites are so far away from the sun that they cannot possibly support any life. It is a certainty, then, that higher forms of life do not exist within our solar system.

DISTANT INHABITED WORLDS

Life may exist on other planets orbiting around the stars and it is conceivable that it has developed to higher forms. Solar systems of our type are probably rather common. We have reason to suppose that about one out of every ten stars has revolving around it a planet with about the same chemical and physical conditions as the earth's. If we were to send a space ship which, like Noah's ark, contained seeds of all plants and one pair of each animal species, to such a planet, a terrestrial community could flourish there too.

If we confine ourselves to our galaxy, we may note that there are about one hundred billion stars in it and that, of the planets in the galaxy, there are perhaps several billion that are inhabitable. But at the present time we have no way of calculating how many of them are actually inhabited. It is possible that all inhabitable planets are sustaining some kind of life; or perhaps only one in every thousand is inhabited; but it is also possible that all of the inhabitable planets are sterile.

INHABITABLE PLANETS AND INHABITED PLANETS

In Chapter One we discussed how life originated on the earth. We have been able to reconstruct the main features of the evolutionary process, but so far we know very little about the details of the complicated series of events which led to the very beginning of life.

Consequently, we are forced to assume that at some point in time, a certain molecular aggregate happened to form that had all the properties—the ability to take up nourishment from its environment, the ability to reproduce itself, and so on—that are necessary for all life. But how large is the probability that this could occur?

It is possible that the process through which life started on earth was completely normal, in the sense that its occurrence was inevitable, given the chemical and physical conditions which prevailed on the earth some billion years ago. For example, to draw an analogy, when the sun shines, water evaporates and later condenses to clouds. Rain falls from the clouds, the water accumulates in various places, and, from these accumulations, rivers and other bodies of fresh water are formed. We can therefore say that the formation of rivers is a necessary and normal consequence of prevailing conditions. Possibly the origin of life, too, was a necessary consequence of a certain set of conditions. If so, then life must have started a thousand times in different places on the earth as soon as conditions became ripe for its outburst.

If we assume that life did indeed begin in this way, we can then conclude that, on those planets which are similar to the earth, the same process has taken place. Consequently life probably exists wherever the conditions are favorable. Every inhabitable planet must be inhabited.

But it is also conceivable that the origin of life on the earth was due to a very unlikely event. An extremely unlikely combination of molecules may have been required, and the probability of that particular combination occurring may have been very close to zero. The origin of life on the earth, if viewed from this perspective, seems to have been almost entirely a result of blind chance. One could almost call it a miracle. If this view is correct, the earth may have been sterile for a long time after it was formed, and life did not originate on the earth because the conditions for it were favorable, but because by chance a remarkable aggregate of molecules was formed.

If we accept this point of view, it follows that not all of the inhabitable planets are inhabited. Perhaps most of them are uninhabited. It is not even impossible that the earth is the only celestial body which accommodates life. Probably most people choose to assume that there are many inhabited planets in our galaxy, but this is no more than a guess. We know too little about the origin of life to be able to deduce its causes with any degree of certainty.

BEINGS EQUIVALENT TO MAN

The next question we must answer is about the origin of man. When life originated, either by chance or as a necessary consequence of the prevailing conditions, biological evolution was set in motion. It in turn has differentiated life into a multitude of different forms. One of the paths of evolution led to the insects, another to the birds, a third to the mammals, and so on.

Were all these courses of evolution necessary? Or were they the result of chance? If life has begun on another planet which is almost exactly like the earth, has the biological evolution there also engendered insects, birds, and mammals? If only insects, but neither birds nor mammals, have evolved, are the insects on that planet more highly developed than they are here? On the earth, there are ants that have formed quite complex social structures. Has some sort of ant on the planet "Xyz" advanced further than the ants on earth and produced a technological culture? Could there possibly be nonmammals of any type who equal, or perhaps even surpass, man in their ability to create culture?

To these questions, too, any answers that we might offer would be only guesses. Our understanding of biological evolution is too scant for us to determine where it is heading. Did man evolve by chance, or as the inevitable consequence of prevailing conditions? We will not be able to answer this question until we have learned much more about the general conditions of life itself.

EQUIVALENT TECHNOLOGY

We do not know, then, whether we have any comrades in the universe. But if we have, do we have any chance of contacting them? It will be a long, long time before we are able to make an odyssey into the infinite depth of space to look for equals. But if a technological culture comparable to ours has arisen anywhere, the possibility of making such contact cannot be disregarded. We have the technological capability to build a transmitter which, by means of radio waves or laser rays, could send signals which could be detected several light years from here provided that a culture possessing sophisticated communications equipment is there to receive the signals. Should our neighbors live ten light years away, it is highly possible that we could receive a reply to our messages twenty years after we had sent them; but should they live one hundred light years away, two hundred years would elapse before we could hope for a reply. It would be a slow and arduous conversation, but no doubt the most interesting that we had ever experienced. If we were to find that somewhere in the universe there was indeed a technological culture equivalent to ours we would immediately know much more about ourselves. First of all such a finding would confirm the theory that life is a normal occurrence and also that biological evolution has a high probability of leading to some type of life form that is at least as highly developed as man. However, besides elucidating some of the most important questions in science, the establishment of such a contact should also clarify central problems in history and sociology.

BASIC QUESTIONS ABOUT CULTURAL EVOLUTION

The discovery of a technological culture elsewhere in the universe would induce us to conclude not only that biological evolution had led to the development of a being comparable to man, but also that sooner or later any such highly developed being would initiate a process of cultural evolution that gave rise to science and technology. To determine whether these deductions are reasonable, we must study a historical-sociological problem of the same type as the problem that arises from biological evolution.

If we suppose that there is a living being possessing the same genetic properties as man, then it is highly probable that he will create a culture. But man's cultural evolution has taken different courses in different parts of the world. Some people became nomads, who foraged, hunted, and fished for their livelihood, whereas others grew their food and developed cultures based on farming. Some cultures became quite advanced, notably the Indian, the Chinese, the Egyptian, the Aztec, and the Incan cultures. The Mediterranean culture, too, became highly civilized and eventually spread across the entire western European continent and beyond. Although contacts between these different cultures were established from time to time and religions and technical inventions spread from one culture to another, the major cultures continued to develop independently of each other for a long time.

Of all of them, it was the Western, or Occidental, culture which developed into a technological culture destined to spread throughout the world, and to invade the other cultural regions. As a result, the global culture is now undergoing its painful birth.

WHY GALILEO?

We have yet a third problem of the same fundamental nature as those on the origin of life and the origin of man. Did the technological culture arise from a necessity or by chance? Both the Chinese and the Indian cultures were in most respects superior to the Western culture as recently as the eighteenth century and perhaps even the beginning of the nineteenth century. It was not until the middle of the nineteenth century that the Western men made the technological leap that gave them the keys to world domination. The reason they were able to make this leap was that as early as the Renaissance they had begun to cultivate science in a way which gradually led to the technological evolution. The one man most responsible for this development was Galileo.

We come now to the central question. Why was there no Chinese or Indian Galileo? During the Tang and Sung periods, culture in China flourished very much as that of the Renaissance in Europe. Not only art and literature but also science and technology thrived as never before. But instead of leading to a scientific-technological break-through, such as that which occurred in the West during the nineteenth century, China's cultural flowering withered, perhaps because, as some have argued, she was burdened with a reactionary bureaucracy. Similar periods of cultural regeneration have occurred in India's history, most recently during the Mogul period.

There is no indication that these cultures could have achieved a scientific-technological breakthrough at any subsequent time. When it did not happen during the Sung or the Mogul eras, the chance for it was apparently irretrievably lost.

It is therefore legitimate to ask how a human culture develops "normally." Was it the Western man who acted normally when he created a technological culture, or does the Chinese-Indian pattern represent a more natural evolution? Indeed a somewhat similar pattern of events took place in the history of the classical Mediterranean culture, which declined once the scientific effort of the Greeks had lost its impetus. To refine the question, let us ask, "Why did Galileo become Galileo?" Was it sheer chance that, at precisely the right time, a man was born with enough intelligence to produce revolutionary ideas and with enough integrity to stand up for them against repression from the church? Suppose that Galileo had died young: would someone else have come forth to start a similar avalanche? Or would the West, like the remains of an imposing ancient forest, have become petrified, its creative scientific potential untapped? And suppose a similar and anonymous man had been born during the Sung or Mogul eras—could he have done what Galileo did?

We must answer this third question if we are to predict our chances of contacting alien civilizations. If life has appeared on another planet, and if a life form equivalent to man has evolved, is it inevitable that these beings will produce a Galileo? Or will our fellows there produce highly civilized cultures that forever remain mute and unknown to us because they have developed no radios and lasers.

A GLOBAL HIROSHIMA?

Finally, a fourth question focuses on the stability of our culture. At present there are a number of destabilizing factors: the population explosion, pollution, and—perhaps the most threatening of all—our rapidly increasing means of destruction (and one must be an optimist indeed to believe that all of this amassed destructive power will always remain in the hands of people who will not use it). We have very good reason to ask if it is our ultimate fate to be our own destroyers. Will we make the earth uninhabitable? If so, will this happen before our culture has established refuge centers outside the earth?

In our investigation of the possibility of other civilizations existing elsewhere in the cosmos, the fourth question is concerned with the life span of a technological culture on another planet. Once established, will such a culture endure indefinitely, or will it annihilate itself so rapidly that we have little hope of contacting it?

Before we can decide whether we are unique in the universe, before we can answer the question of whether there are foreign cultures for us to contact, we must answer four fundamental questions about ourselves: Is life a result of chance? Is man a result of chance? Was Galileo a result of chance? Is a global Hiroshima unavoidable? Thus we find that the question whether we are unique is closely connected with the destiny of life on the third planet.

Are we a "normal" phenomenon? When, in the great galactic factory, the solar systems were manufactured on a production belt, was a cybernetic culture for the third planet automatically produced as standard equipment? Does the comfortable life many people live in the industrialized countries constitute a "normal" result of the interplay of cosmic, biological, and technological-sociological factors, a type of life which eventually all people in the world will enjoy forever?

Alternatively, do we constitute an extremely rare error in production? Are we the result of three highly unlikely events, enjoying a precarious existence in a world of threats that we ourselves have created as we continue our diligent efforts to learn how to destroy ourselves?

We do not know which of these designs applies to life on the third planet. But in spite of our ignorance of mankind's origin and purpose, it is essential that we clarify our present situation, that we devise a strategy which may perhaps give us a chance to master our destiny.

XII

References and Comments

Instead of listing bibliographic references in the text itself, we have chosen to include them in this appendix, along with comments on some of the material that we have used.

Some of the ideas developed in Chapters One, Two, Three, and Eleven are more extensively explored in Hannes Alfvén, *Atom, Man, and the Universe: The Long Chain of Complications*, San Francisco, W. H. Freeman and Company, 1969.

In Chapters Four through Ten, we have as a rule avoided using examples from today's politics. Instead we have chosen older and therefore less controversial examples in order to elucidate modern conditions which are not fundamentally unlike those which the examples illustrate. We have not used examples from South America and Africa because we have never had an opportunity to visit these continents.

CHAPTER I

J. B. Calhoun, "Population Density and Social Pathology," *Scientific American*, **206**, Feb. 1962, p.139. Available as *Scientific American* Offprint No. 506 from W. H. Freeman and Company.

J. B. Calhoun, "A Method for Self-Control of Population Growth Among Mammals," *Science*, **109**, 1949, p. 333.

V. C. Wynne-Edwards, "Population Control in Animals," *Scientific American*, **211**, Aug. 1964, p. 68. Available as *Scientific American* Offprint No. 192 from W. H. Freeman and Company.

CHAPTER II

Desmond Morris, *The Naked Apes: A Zoologist's Study of the Human Animal*, London, 1967; New York, McGraw-Hill, 1968.

CHAPTER III

Peter J. Glaser, "Power from the Sun," *Science*, **162**, 1968, p. 857.

Freeman J. Dyson, "Interstellar Transport," *Physics Today*, **21**, Oct. 1958, p. 41.

CHAPTER V

Aggression as a Biological Inheritance. The question of whether war is caused by an innate aggressive instinct in man has given rise to much debate. Konrad Lorenz, in several books (especially *On Aggression*, New York, Harcourt Brace

& World, 1966), has described the aggression of different animal species and maintained that man has an inherited aggressive instinct which would largely account for wars. His view has been sharply criticized by fourteen leading biologists and anthropologists who have jointly assembled their arguments in *Man and Aggression*, edited by M. F. Ashley Montagu, New York, Oxford University Press, 1968. According to the authors of these articles, Lorenz's conclusions were based on the study of animals in captivity and may have little relevance to the way animals live in a natural state. Although he is a recognized authority in his special field, Lorenz's views are considered obsolete in the wider area that he is dealing with. Two of the essays in the Montagu collection are entitled: "Man Has No Killer Instinct" and "War Is Not in Our Genes." The book should be read by all who have read Lorenz, and by others too. See also Ashley Montagu, "Animals and Men: Divergent Behavior," *Science*, 61, 1968, p. 963. Other important contributions to the discussion include John D. Carthy and Francis J. Ebling, Eds., *The Natural History of Aggression*, New York, Academic Press, 1965, and a severely critical review of it by Anatol Rapoport, "Is Warmaking a Characteristic of Human Beings or of Culture?" *Scientific American*, 213, Oct. 1965, p. 115. Another work we have consulted is N. Tinbergen, "On War and Peace in Animals and Men," *Science*, 160, June 1968, p. 1411.

War as a Social Enterprise. See the following works for further information on Japan's military history.

Ienaga Saburo, *History of Japan*, 5th Edition, Tokyo, Japan Travel Bureau, 1961.

Kenneth Scott Latourette, *The History of Japan*, New York, Macmillan Co., 1957.

Hugh Borton, *Japan's Modern Century*, New York, Ronald Press, 1955.

CHAPTER VI

In Chapter Six, we have introduced somewhat stereotyped portraits of the general, the politician, the businessman, and the scientist. One may raise objections to these generalizations. Perhaps many would desire more precise distinctions between "nice" politicians—that is, those they vote for—and "nasty" politicians—those who are supposedly responsible for our troubles. The purpose of these brief descriptions is not to air our personal opinions but only to capture a leader's general behavior pattern when he is playing his particular role in the social structure. We have sought simply to stress the fact that, at present, neither politicians nor any of the other groups of leaders are able to direct their efforts to solving the problems of global coexistence and thereby ward off threatening catastrophes.

Incompetent Power Groups. The relationship between Clemenceau and the generals during the different phases of the war was of course more complicated. Some of his biographers do not mention the familiar quotation we have cited, while others insist the remark was made on an earlier occasion. See Norbert Guterman, *A Book of French Quotations*, Garden City, N.Y., Doubleday, 1963. The remark is also ascribed to Talleyrand and Briand.

CHAPTER VII

The Increasing Illiteracy. Forty-four percent of the world's population older than fifteen years of age was illiterate in 1950, and this figure had decreased to thirty-nine percent in 1968. However, the population of the world increased from 2.5 billion to 3.5 billion, and consequently the number of illiterates in 1968 was nearly 300 million more than in 1950.

See *UNESCO Basic Facts and Figures 1952–1961,*

UNESCO Statistical Yearbook 1963, and *The World Population Data Sheet*, 1968, a yearly information sheet published by the Population Reference Bureau, Washington, D.C. The procedures used to eliminate illiteracy in the Soviet Union are documented in Mikhail N. Zinov'ev and Aleksandra V. Pleshakova, *How Illiteracy Was Wiped Out in the USSR*, Moscow, Foreign Languages Publishing House, 1961. Although this arresting, short book may not be entirely free of propaganda, there is little reason to doubt the major events in the account. It is recommended for those who work with similar problems in underdeveloped countries.

Sir Charles Jeffries, in *Illiteracy: a World Problem*, New York, Praeger, 1967, describes with deep involvement the problem of illiteracy. In an article entitled "International Year for Human Rights," *UNESCO Courier*, April 1968, p. 33, one finds the following: "In a world where vast areas are still in the grip of hunger and malnutrition, and where more than 700,000,000 illiterate persons are shut out from the world of ideas mirrored in the written word, there are many people for whom the provisions of the Universal Declaration are nothing more than empty promises."

The Ignorance of the Rulers. The history of the Maginot Line is covered in a number of books whose titles alone convey the hopes and delusions inspired by the "impregnable" defense. Some examples are the following:

Vivian Rowe, *The Great Wall, the Triumph of the Maginot Line*, New York, Putnam, 1961.

Eitenne Anthereu, *Grandeur et Sacrifice de la Ligne Maginot*, Paris, G. Durassie et Cie. Editeur, 1962.

Géneral Pretelat, *Le Destin Tragique de la Ligne Maginot*, Paris, Edition Berger-Levrault, 1950.

The case of the Maginot Line is by no means unique. In the early days of World War II, the defense of the British

Empire in the Far East was largely based on the assumption that Singapore was impregnable. When the Japanese attacked Singapore, it was found that the city was vulnerable to a land-based assault. In writing about this episode, Winston Churchill commented that he ought to have known about Singapore's weakness, his advisors ought to have known about it, he ought to have been informed, and he ought to have asked about it. The reason he did not look into this potential danger, along with the multitude of other problems he was dealing with, was that he could no more imagine the possibility that Singapore could be taken in a land attack than he could imagine that a battleship could be launched without a bottom.

Analogous incidents are probably just as common today—not only in wartime but during peace as well. But we do not recognize them for the dangers they present until enough time has passed for the threat to subside. For example, how do we know if there is a safety latch on the hydrogen bomb button, or if the cork in the bottle of bacterial warfare is tight?

Government and Judgment. Facets of Chamberlain's fateful mission at Munich may be investigated by consulting the following works:

Neville Chamberlain, *In Search of Peace*, New York, Putnam, 1939.

Stuart Hodgson, *The Man Who Made the Peace: Neville Chamberlain*, New York, Dutton, 1938.

Keith Feiling, *Life of Neville Chamberlain*, London, Macmillan, 1947.

Our Ignorance About Each Other. Joseph Needham's *Science and Civilization in China*, Cambridge, England, Cambridge University Press, 1956, is a work that is difficult to read, but it is powerful enough to shake one's vision of world history. Many of the "world history" courses offered at universities and high schools in the Western part of the world are responsible for grotesque distortions of history because they concentrate almost exclusively on the Hellenic-Judaic-Christian cultural tradition.

CHAPTER IX

The Increase in Population. The question of how large a group is needed to sustain a culture is very interesting. The Greek cultural explosion occurred within a very small group. During the Golden Age of Athens the number of free men probably did not exceed 5,000. The total population of Attica, which labored so that the privileged citizens of Athens could prosper, was not larger than a few hundred thousand. The Icelanders offer another example of a small group of people (at times less than 50,000, now approximately 200,000) who, despite their isolation and periodically severe poverty, have honorably sustained a distinguished culture of their own.

In many outlines of the problems in underdeveloped countries, these nations are grouped together and the reader is led to believe that the conditions are similar in them all. But such a view is grossly misleading, especially if China is categorized as an underdeveloped country. The often quoted statement that two-thirds of the world's population lives in underdeveloped countries at or below the starvation level, presupposes that China is one of these countries. Though China did suffer a famine in about 1960, when the harvests were very poor, it has since enjoyed a number of good harvests and the food supply there is probably quite good. If the 800 million inhabitants of China can be said to be reasonably well nourished, then it is less than half of the world population that lives continually under the threat of starvation—still a sufficiently shocking number.

It is also misleading to liken the problems in China to those of India, Africa, and South America, as some critics have done. It cannot be said, in fact, that a single all-inclusive "underdeveloped country" problem exists. There are at least three: an Indian-Pakistani problem, an African problem, and a South American problem. All should probably be approached and solved in quite different ways.

Technical and Political Possibilities. Small-scale experiments conducted in India and Pakistan have indicated that sterilization by surgery may become an important aid to birth control.

The Japanese Example. The population growth in Japan is well documented by census surveys that have been conducted every fifth year since 1920, when the birth rate was about 35 per thousand, and the mortality rate 23 per thousand. The mortality rate decreased little by little until the end of the 1940's, when it rapidly declined from 16.8 to 9.4 per thousand. The birth rate tended to decrease slightly until 1940, when this tendency was reversed so that in 1947 it attained a temporary peak of 34.3 per thousand. After 1947, the birth rate decreased rapidly so that in 1957 it had shrunk to 17.3 per thousand. Since then it has fluctuated around this low number. What happened in Japan? In 1945 the population of Japan was 72 million, and 3 years later it had grown to 80 million. Part of the population increase was due to the return of Japanese from the territories lost in the war, but it was primarily due to the baby boom that followed the demobilization. But the defeat had hit the Japanese people hard, and the domestic economy was largely shattered, especially in the big cities. Faced with economic shortage, the Japanese soon realized that the birth rate had to be decreased if the standard of living were not to sink even further. The press and the radio stations took the lead in attacking the population problem, and they are acknowledged to have had a great role in popularizing birth control. Protests against birth control came from religious and other ideological sources, but they were largely ineffective in the face of the stronger wish of the majority of the people to decrease the number of children.

Japanese women who did not wish to have children had to resort to abortions in spite of the fact that they had been forbidden by law since 1940. The abortion law that had forced women to resort to illegal abortions appeared obsolete, and the Japanese medical association called a meeting in 1947 to open discussions about it. These discussions went on and, without any serious resistance, resulted in the decision by the Japanese parliament to introduce a new law, the so-called "eugenic protection law." The law had two purposes: first to eliminate the criminal abortions and their harmful consequences, and second to stop the shockingly rapid population increase.

But of course the problem was not that easily solved. Resorting to legalized abortions did not appear ideal. But the government found it difficult to insist upon further measures. By 1949, however, it found it necessary to authorize the sale of contraceptives. Not until 1952 did the state health care authority become responsible for a family planning program.

The following three-point program was initiated: (1) general propaganda in support of family planning; (2) distribution of information to groups; (3) distribution of information to individuals, including detailed advice about contraceptives. Doctors, district nurses, and midwives actually served as instructors, especially in their personal contacts with the people. But for geographical and other practical reasons, much of the responsibility was assumed by the midwives. Instructional courses were arranged for them, and upon completing these, those who qualified were given the title of "family planning instructor," which also gave them the right to sell contraceptives. The instructors were encouraged to seek out clients in the villages personally.

In spite of these measures, the number of abortions continued to rise until 1955, when the annual figures showed that 1.17 million had occurred. As a consequence, the campaign for family planning was reinforced. It is of special interest to note that many industrial firms had started private campaigns in their establishments and had integrated family planning into their social services. In three years this drive boosted the use of contraceptive methods by the total population from forty percent to seventy percent.

The Japanese example is well worth studying, even if it is not fully applicable to those populous nations which have not yet taken even the problem of illiteracy seriously. Japan, as an advanced industrialized nation, had a literate population and, in spite of its proximity to Asia, its demographic structure was more similar to that of the West than to the East's. Also, according to some reports, Japan has had a long history of practicing abortion as a birth-control method.

The Japanese measures for regulating the population are of great importance to everyone. Thus it is surprising that the details about them are so little known, and that they have received so little attention from those responsible for family planning programs in underdeveloped countries. Even in modern history books about Japan, these efforts are hardly mentioned.

The above information is taken largely from a lecture by Minoru Muramatsu, M.D., Institute of Public Health in Japan. It is published in B. R. Berelson et al., Eds., *Family Planning and Population Programs: A Review of World Development*, Chicago, University of Chicago Press, 1966. See also the yearbook *Nippon, a Chartered Survey of Japan*, Tokyo, 1968.

CHAPTER XI

See Walter Sullivan, *We Are Not Alone*, New York, McGraw-Hill, 1964. Sullivan maintains that we can assert with certainty the existence of alien civilizations. The polemic in Chapter Eleven is in part an argument against this opinion. See also A. G. Cameron, Ed., *Interstellar Communication*, Menlo Park, Calif., Benjamin, Inc., 1963.